Advanced R 4 Data Programming and the Cloud

Using PostgreSQL, AWS, and Shiny

Second Edition

Matt Wiley
Joshua F. Wiley

Apress®

Advanced R 4 Data Programming and the Cloud: Using PostgreSQL, AWS, and Shiny

Matt Wiley
Victoria College
Victoria, TX, USA

Joshua F. Wiley
Monash University
Melbourne, VIC, Australia

ISBN-13 (pbk): 978-1-4842-5972-6
https://doi.org/10.1007/978-1-4842-5973-3

ISBN-13 (electronic): 978-1-4842-5973-3

Managing Director, Apress Media LLC: Welmoed Spahr
Acquisitions Editor: Steve Anglin
Development Editor: Matthew Moodie
Editorial Operations Manager: Mark Powers

Cover designed by eStudioCalamar
Cover image by Cris-Dinoto on Unsplash (www.unsplash.com)

Distributed to the book trade worldwide by Apress Media, LLC, 1 New York Plaza, New York, NY 10004, U.S.A. Phone 1-800-SPRINGER, fax (201) 348-4505, e-mail orders-ny@springer-sbm.com, or visit www. springeronline.com. Apress Media, LLC is a California LLC and the sole member (owner) is Springer Science + Business Media Finance Inc (SSBM Finance Inc). SSBM Finance Inc is a **Delaware** corporation.

For information on translations, please e-mail editorial@apress.com; for reprint, paperback, or audio rights, please email bookpermissions@springernature.com.

Apress titles may be purchased in bulk for academic, corporate, or promotional use. eBook versions and licenses are also available for most titles. For more information, reference our Print and eBook Bulk Sales web page at http://www.apress.com/bulk-sales.

Any source code or other supplementary material referenced by the author in this book is available to readers on GitHub via the book's product page, located at www.apress.com/9781484259726. For more detailed information, please visit http://www.apress.com/source-code.

Printed on acid-free paper

Table of Contents

About the Authors

Matt Wiley leads institutional effectiveness, research, and assessment at Victoria College, facilitating strategic and unit planning, data-informed decision making, and state/regional/federal accountability. As a tenured, associate professor of mathematics, he won awards in both mathematics education (California) and student engagement (Texas). Matt earned degrees in computer science, business, and pure mathematics from the University of California and Texas A&M systems.

Outside academia, he coauthors books about the popular R programming language and was managing partner of a statistical consultancy for almost a decade. He has programming experience with R, SQL, C++, Ruby, Fortran, and JavaScript.

A programmer, a published author, a mathematician, and a transformational leader, Matt has always melded his passion for writing with his joy of logical problem solving and data science. From the boardroom to the classroom, he enjoys finding dynamic ways to partner with interdisciplinary and diverse teams to make complex ideas and projects understandable and solvable. Matt enjoys being found online via Twitter @matt_math or http://mattwiley.org/.

Joshua F. Wiley is a lecturer in the Turner Institute for Brain and Mental Health and School of Psychological Sciences at Monash University. He earned his PhD from the University of California, Los Angeles, and completed his postdoctoral training in primary care and prevention. His research uses advanced quantitative methods to understand the dynamics between psychosocial factors, sleep, and other health behaviors in relation to psychological and physical health. He develops or codevelops a number of

R packages including *varian*, a package to conduct Bayesian scale-location structural equation models; *MplusAutomation*, a popular package that links R to the commercial Mplus software; *extraoperators* for faster logical operations; `multilevelTools` for diagnostics, effect sizes, and easy display of multilevel/mixed-effects models results; and miscellaneous functions to explore data or speed up analysis in *JWileymisc*. Joshua enjoys being found online via Twitter @WileyResearch or `http://joshuawiley.com/`.

About the Technical Reviewer

 **Andrew Moskowitz** is an analytics and data science professional in the entertainment industry focused on understanding user behavior, marketing attribution, and efficacy and using advanced data science concepts to address business problems. He earned his PhD in quantitative psychology at the University of California, Los Angeles, where he focused on hypothesis testing and mixed-effects models.

Acknowledgments

The authors would like to thank the vibrant R community for their hard work, their cleverness, their willingness to create and share, and most importantly their *kindness* to new learners. One author remembers first getting into programming (in a different language than R) and very early on trying to use an early incarnation of search engine to discover what "rtfm" meant. One of R's **best** community features is a willingness to share and not mistake *new learner* for *someone to be "hazed."* This civility is something to nurture and cherish.

> *...an elegant weapon for a more civilized age.*
>
> —Ben Kenobi

We would also like to thank Andrew for sticking with us for a second edition and a bleeding-edge version release of R. Thank you, Andrew.

Last, and not least, we would like to thank our family. You continue to support us through a lot of late evenings, early mornings, and the times in between. Thank you.

CHAPTER 1

Programming Basics

As with most languages, becoming a power user requires extra understanding of the underlying structure or rules. Data science through R is powerful, and this chapter discusses such programming basics including **objects**, **operators**, and **functions**.

Before we dig too deeply into R, some general principles to follow may well be in order. First, experimentation is good. It is much more powerful to learn hands-on than it is simply to read. Download the source files that come with this text, and try new things!

Second, it can help quite a bit to become familiar with the ? function. Simply type ? immediately followed by text in your R console to call up help of some kind—for example, ?sum. We cover more on functions later, but this is too useful to ignore until that time. Using your favorite search engine is also wise (such as this search string: R sum na). While memorizing some things may be helpful, much of programming is gaining skill with effective search.

Finally, just before we dive into the real reason you bought this book, a word of caution: this is an applied text. Our goal is to get you up and running as quickly as possible toward some useful skills. A rigorous treatment of most of these topics—even or especially the ideas in this chapter—is worthwhile, yet beyond the scope of this book.

1.1 Software Choices and Reproducibility

This book is written for more experienced users of the R language, and we suppose readers are familiar with installing R. For completeness, we list the primary software used throughout this book in Table 1-1. Individual R packages will be introduced inside any chapter where their use is indicated. Specifics of setting up an Amazon cloud compute instance will be walked through in the relevant cloud computing chapter as well.

© Matt Wiley and Joshua F. Wiley 2020
M. Wiley and J. F. Wiley, *Advanced R 4 Data Programming and the Cloud*,
https://doi.org/10.1007/978-1-4842-5973-3_1

Table 1-1. *Advanced R Tech Stack*

Software	URL
R 4.0.0	https://cran.r-project.org/
RStudio 1.3.959	https://rstudio.com/products/rstudio/download/
Windows 10	www.microsoft.com/
Java 14	www.oracle.com/technetwork/java/javase/overview/index.html
Amazon Cloud	https://aws.amazon.com/
Ubuntu	https://ubuntu.com/download

For a complete walk-through of how to install R and RStudio on Windows or Macintosh, please see our *Beginning R* [37] book.

1.2 Reproducing Results

One useful feature of R is the abundance of packages written by experts worldwide. This is also potentially the Achilles' heel of using R: from the version of R itself to the version of particular packages, lots of code specifics are in flux. Your code has the potential to not work from day to day, let alone our code written weeks or months before this book was published.

All code used in the following chapters will be hosted on GitHub. Code there may well be more recent than printed in this text. Should code in the text not work due to package changes or base R changes, please visit this book's GitHub site:

```
options(
  width = 70,
  stringsAsFactors = FALSE,
  digits = 2)
```

1.3 Types of Objects

First of all, we need things to build our language, and in R, these are called objects. We start with five very common types of objects.

Logical objects take on just two values: TRUE or FALSE. Computers are binary machines, and data often may be recorded and modeled in an all-or-nothing world. These logical values can be helpful, where TRUE has a value of 1 and FALSE has a value of 0.

As a reminder, # (e.g., the pound sign or hashtag) is an indicator of a code *comment*. The words that follow the # are not processed by R and are meant to help the reader:

```
TRUE ## logical
```

```
## [1] TRUE
```

```
FALSE ## logical
```

```
## [1] FALSE
```

As you may remember from some quickly muttered comments of your college algebra professor, there are many types of numbers. Whole numbers, which include zero as well as negative values, are called *integers*. In set notation, ... ,-2, -1, 0, 1, 2, ... , these numbers are useful for headcounts or other indexes. In R, integers have a capital L suffix. If decimal numbers are needed, then double numeric objects are in order. These are the numbers suited for ratio data types. Complex numbers have useful properties as well and are understood precisely as you might expect, with an i suffix on the imaginary portion. R is quite friendly in using all of these numbers, and you simply type in the desired numbers (remember to add the L or i suffix as needed):

```
42L ## integer
```

```
## [1] 42
```

```
1.5 ## double numeric
```

```
## [1] 1.5
```

```
2+3i ## complex number
```

```
## [1] 2+3i
```

Nominal-level data may be stored via the character class and is designated with quotation marks:

```
"a" ## character
```

```
## [1] "a"
```

Of course, numerical data may have missing values. These missing values are of the type that the rest of the data in that set would be (we discuss data storage shortly). Nevertheless, it can be helpful to know how to hand-code logical, integer, double, complex, or character missing values:

```
NA ## logical
```

```
## [1] NA
```

```
NA_integer_ ## integer
```

```
## [1] NA
```

```
NA_real_ ## double / numeric
```

```
## [1] NA
```

```
NA_character_ ## character
```

```
## [1] NA
```

```
NA_complex_ ## complex
```

```
## [1] NA
```

Factors are a special kind of object, not so useful for general programming, but used a fair amount in statistics. A factor variable indicates that a variable should be treated discretely. Factors are stored as integers, with labels to indicate the original value:

```
factor(1:3)
```

```
## [1] 1 2 3
## Levels: 1 2 3
```

```
factor(c("alice", "bob", "charlie"))
```

```
## [1] alice    bob    charlie
## Levels: alice bob charlie
```

```
factor(letters[1:3])
```

```
## [1] a b c
## Levels: a b c
```

We turn now to data structures, which can store objects of the types we have discussed (and of course more). A *vector* is a relatively simple data storage object. A simple way to create a vector is with the concatenate function `c()`:

```
## vector
c(1, 2, 3)

## [1] 1 2 3
```

Just as in mathematics, a *scalar* is a vector of just length 1. Toward the opposite end of the continuum, a *matrix* is a vector with dimensions for both rows and columns. Notice the way the matrix is populated with the numbers 1-6, counting down each column:

```
## scalar is just a vector of length one
c(1)

## [1] 1

## matrix is a vector with dimensions
matrix(c(1:6), nrow = 3, ncol = 2)

##      [,1] [,2]
## [1,]    1    4
## [2,]    2    5
## [3,]    3    6
```

All vectors, be they scalar, vector, or matrix, can have only one data type (e.g., integer, logical, or complex). If more than one type of data is needed, it may make sense to store the data in a list. A list is a vector of objects, in which each element of the list may be a different type. In the following example, we build a list that has character, vector, and matrix elements:

```
## vectors and matrices can only have one type of data (e.g., integer,
## logical, etc.)

## list is a vector of objects
## lists can have different type of objects in each element
list(
  c("a"),
```

```
  c(1, 2, 3),
  matrix(c(1:6), nrow = 3, ncol = 2)
)
```

```
## [[1]]
## [1] "a"
##
## [[2]]
## [1] 1 2 3
##
## [[3]]
##      [,1] [,2]
## [1,]    1    4
## [2,]    2    5
## [3,]    3    6
```

A particular type of list is the *data frame,* in which each element of the list is identical in length (although not necessarily in object type). With the underlying building blocks of the simpler objects, more complex structures evolve. Take a look at the following instructive examples with output:

```
## data frames are special type of lists
## where each element of the list is identical in length
data.frame(
  1:3,
  4:6)
```

```
##   X1.3 X4.6
## 1    1    4
## 2    2    5
## 3    3    6
```

```
## using non equal length objects causes problems
data.frame(
  1:3,
  4:5)
```

```
## Error in data.frame(1:3, 4:5): arguments imply differing number of
rows: 3, 2

data.frame( 1:3, letters[1:3])

##   X1.3 letters.1.3.
## 1    1             a
## 2    2             b
## 3    3             c
```

Because of their superior computational speed, in this text we primarily use *data table* objects in R from the data.table package [9]. Data tables are similar to data frames, yet are designed to be more memory efficient and faster (mostly due to more underlying C++ code). Even though we recommend data tables, we show some examples with data frames as well because when working with R, much historical code includes data frames and indeed data tables inherit many methods from data frames (notice the last line of code that follows shows TRUE):

```
##if not yet installed, run the below line of code if needed.
#install.packages("data.table")
library(data.table)
## data.table 1.12.8 using 6 threads (see ?getDTthreads). Latest news:
r-datatable.com

dataTable <- data.table( 1:3, 4:6)
dataTable

##    V1 V2
## 1:  1  4
## 2:  2  5
## 3:  3  6

is.data.frame(dataTable)

## [1] TRUE
```

It is worth mentioning at this stage a little bit about the data structure "wars." Historically, the predominant way to structure the types of row/column or *tabular* data many researchers use was data *frames*. As data grew in column width and row length, this base R structure no longer solved everyone's needs. Grown out of the same SQL data control mindset as some of the largest databases, the `data.table` package/library is (these days) suited to multiple computer cores and efficient memory operations and uses more-efficient-than-R languages under the hood. For the largest data sets and for those who have any background in SQL or other programming languages, data *tables* are hugely effective and intuitive. Not all folks first coming to R have a programming background (and indeed that is a very good thing). A competing data structure, the *tibble*, is part of what is called the *tidyverse* (a portmanteau of tidy and universe). Tibbles, like data tables, are also data frames at heart, yet they are improved. In the authors' opinion, while not yet quite as fast as tables, tibbles have a more new-user-friendly style or language syntax. They're beloved by a large part of the R community and are an important part of modern R. In practice, data tables are still faster and can often achieve tasks your authors find most common in fewer lines of code. Both these newer structures have their strengths, and both have their place in the R universe (and indeed Chapters 7 and 8 focus on data tables, yet time is given to tibbles in Chapter 9). All the same, this text will primarily use data tables.

Having explored several types of objects, we turn our attention to ways of manipulating those objects with operators and functions.

1.4 Base Operators and Functions

Objects are not enough for a language; while nouns are nice, actions are required. Operators and functions are the verbs of the programming world. We start with assignment, which can be done in two ways. Much like written languages, more elegant turns of phrase can be more helpful than simpler prose. So although both = and <- are assignment operators and do the same thing, because = is used within functions to set arguments, we recommend for clarity's sake to use <- for general assignment. We nevertheless demonstrate both assignment techniques. Assignments allow objects to be given sensible names; this can significantly enhance code readability (for your future self as well as for other users).

In addition to assigning names to variables, you can check specifics by using functions. Functions in R take the general format of function name, followed by parentheses, with input inside the parentheses, and then R provides output. Here are examples:

```
x <- 5
y = 3
x

## [1] 5

y

## [1] 3

is.integer(x)

## [1] FALSE

is.double(y)

## [1] TRUE

is.vector(x)

## [1] TRUE
```

It can help to be able to pronounce and speak these lines of code (and indeed that idea is at the heart of our preference for <- which reads "is assigned"). The preceding code might well be read "the variable x *is assigned* the integer value 5." Contrastingly, while the precise same under-the-hood operation is occurring in the next line with y, saying "y *equals* 3" is perhaps less clear as to whether we are discussing an innate property of y vs. performing an assignment.

Once an object is assigned, you can access specific object elements by using brackets. Most computer languages start their indexing at either 0 or 1. R starts indexing at 1. Also, note you can readily change old assignments with little trouble and no warning; it is wise to watch names cautiously and comment code carefully:

```
x <- c("a", "b", "c")
x[1]

## [1] "a"
```

```
is.vector(x)
```

```
## [1] TRUE
```

```
is.vector(x[1])
```

```
## [1] TRUE
```

```
is.character(x[1])
```

```
## [1] TRUE
```

What do we mean by "watch names carefully"? We called the preceding vector x, and it was not a very interesting name. Tough to remember x has the first three letters of the alphabet. Instead, we might choose a better variable name, swapping the tough-to-recall x with passingLetterGrades. Even better, ours makes sense when spoken variable can easily be improved later, such as if we wanted to add the sometimes passing letter grade D or maybe the pass/fail passing grade S:

```
passingLetterGrades <- c("A", "B", "C")
passingLetterGrades[2]
```

```
## [1] "B"
```

While a vector may take only a single index, more complex structures require more indices. For the matrix you met earlier, the first index is the row, and the second is for column position. Notice that after building a matrix and assigning it, there are many ways to access various combinations of elements. This process of accessing just some of the elements is sometimes called *subsetting*:

```
x2 <- matrix(c(1, 2, 3, 4, 5, 6), nrow = 3, ncol = 2)
x2 ## print to see full matrix
```

```
##      [,1] [,2]
## [1,]    1    4
## [2,]    2    5
## [3,]    3    6
```

```
x2[1, 2] ## row 1, column 2
```

```
## [1] 4
```

```
x2[1, ] ## all row 1
```

```
## [1] 1 4
```

```
x2[, 1] ## all column 1
```

```
## [1] 1 2 3
```

```
## can also grab several at once
x2[c(1, 2), ] ## rows 1 and 2
```

```
##      [,1] [,2]
## [1,]   1    4
## [2,]   2    5
```

```
x2[c(1, 3), ] ## rows 1 and 3
```

```
##      [,1] [,2]
## [1,]   1    4
## [2,]   3    6
```

```
## can drop one element using negative values
x[-2] ## drop element two
```

```
## [1] "a" "c"
```

```
x2[, -2] ## drop column two
```

```
## [1] 1 2 3
```

```
x2[-1, ] ## drop row 1
```

```
##      [,1] [,2]
## [1,]   2    5
## [2,]   3    6
```

```
is.vector(x2)
```

```
## [1] FALSE
```

```
is.matrix(x2)
```

```
## [1] TRUE
```

Accessing and subsetting lists is perhaps a trifle more complex, yet all the more essential to learn and master for later techniques. A single index in a single bracket returns the entire element at that spot (recall that for a list, each element may be a vector or just a single object). Using double brackets returns the object within that element of the list—nothing more.

Thus, the following code is, in fact, a vector with the element a inside. Again, using the data-type-checking functions can be helpful in learning how to interpret various pieces of code:

```
## for lists using a single bracket
## returns a list with one element
y <- list(
  c("a"),
  c(1:3))

y[1]

## [[1]]
## [1] "a"

is.vector(y[1])

## [1] TRUE

is.list(y[1])

## [1] TRUE

is.character(y[1])

## [1] FALSE
```

Contrast that with this code, which is simply the element a:

```
## using double bracket returns the object within that
## element of the list, nothing more
y[[1]]

## [1] "a"

is.vector(y[[1]])
```

```
## [1] TRUE
```

```
is.list(y[[1]])
```

```
## [1] FALSE
```

```
is.character(y[[1]])
```

```
## [1] TRUE
```

You can, in fact, chain brackets together, so the second element of the list (a vector with the numbers 1–3) can be accessed, and then, within that vector, the third element can be accessed:

```
## can chain brackets togeter
y[[2]][3] ## second element of the list, third element of the vector
```

```
## [1] 3
```

Brackets almost always work, depending on the type of object, but there may be additional ways to access components. Named data frames and lists can use the $ operator. Notice in the following code how the bracket or dollar sign ends up being equivalent:

```
x3 <- data.frame(
  A = 1:3,
  B = 4:6)
```

```
y2 <- list(
  C = c("a"),
  D = c(1, 2, 3))
```

```
x3$A
```

```
## [1] 1 2 3
```

```
y2$C
```

```
## [1] "a"
```

```
## these are equivalent to
x3[["A"]]
```

```
## [1] 1 2 3
```

```
y2[["C"]]
```

```
## [1] "a"
```

Notice that although both data frames and lists are lists, neither is a matrix:

```
is.list(x3)
```

```
## [1] TRUE
```

```
is.list(y2)
```

```
## [1] TRUE
```

```
is.matrix(x3)
```

```
## [1] FALSE
```

```
is.matrix(y2)
```

```
## [1] FALSE
```

Moreover, despite not being matrices, because of their special nature (i.e., all elements have equal length), data frames and data tables can be indexed similarly to matrices:

```
x3[1, 1]
```

```
## [1] 1
```

```
x3[1, ]
```

```
## A B
## 1 1 4
```

```
x3[, 1]
```

```
## [1] 1 2 3
```

Any named object can be indexed by using the names rather than the positional numbers, provided those names have been set:

```
x3[1, "A"]
```

```
## [1] 1
```

```
x3[, "A"]
```

```
## [1] 1 2 3
```

For data frames, this applies to both column and row names, and these names can be established after building the matrix:

```
rownames(x3) <- c("first", "second", "third")
```

```
x3["second", "B"]
```

```
## [1] 5
```

Data tables use a slightly different approach. While we will devote two later chapters to using data tables, for now, we mention a few facts. Selecting rows works almost identically, but selecting columns does not require quotes. Additionally, you can select multiples by name without quotes by using the .() operator. Should you need to use quotes, the data table can be accessed by using the option with = FALSE such as follows:

```
x4 <- data.table(
  A = 1:3,
  B = 4:6)
```

```
x4[1, ]
```

```
##    A B
## 1: 1 4
```

```
x4[, A] #no quote needed for a column in data.table
```

```
## [1] 1 2 3
```

```
x4[1, A]
```

```
## [1] 1
```

```
x4[1:2, .(A, B)]
```

```
##    A B
## 1: 1 4
## 2: 2 5
```

```
x4[1, "A", with = FALSE]
```

```
##    A
## 1: 1
```

Remember, we said that everything in R is either an object or a function. Those are the two building blocks. So, technically, the bracket operators are functions. Although they're not used as functions with the telltale parens (), they can be. Most functions are named, but the brackets are a particular case and require using single quotes in the regular function format, as in the following example:

```
'['(x, 1)
```

```
## [1] "a"
```

```
'['(x3, "second", "A")
```

```
## [1] 2
```

```
'[['(y, 2)
```

```
## [1] 1 2 3
```

In practice of course, this is almost never used this way. It only needs saying to understand you can code up your own meaning for any function (we devote a chapter on writing your own functions). In fact, this is conceptually how data table gets away with not using quotes for the column names—it changed the way the bracket function works when is.data.table() returns TRUE.

Although we have been using the is.datatype() function to better illustrate what an object is, you can do more. Specifically, you can check whether a value is missing an element by using the is.na() function:

```
NA == NA ## does not work
```

```
## [1] NA
```

```
is.na(NA) ## works
```

```
## [1] TRUE
```

Of course, the preceding code snippet usually has a vector or matrix element argument whose populated status is up for debate. Our last (for now) exploratory function is the inherits() function. It is helpful when no is.class() function exists, which can occur when specific classes outside the core ones you have seen presented so far are developed:

```
inherits(x3, "data.frame")
```

```
## [1] TRUE
```

```
inherits(x2, "matrix")
```

```
## [1] TRUE
```

You can also force lower types into higher types. This coercion can be helpful but may have unintended consequences. It can be particularly risky if you have a more advanced data object being coerced to a lesser type (pay close attention to the attempt to coerce an integer):

```
as.integer(3.8)
```

```
## [1] 3
```

```
as.character(3)
```

```
## [1] "3"
```

```
as.numeric(3)
```

```
## [1] 3
```

```
as.complex(3)
```

```
## [1] 3+0i
```

```
as.factor(3)
```

```
## [1] 3
## Levels: 3
```

```
as.matrix(3)
```

```
## [,1]
## [1,] 3
```

```
as.data.frame(3)
```

```
## 3
## 1 3
```

```
as.list(3)
```

```
## [[1]]
## [1] 3
```

```
as.logical("a") ## NA no warning
```

```
## [1] NA
```

```
as.logical(3) ## TRUE, no warning
```

```
## [1] TRUE
```

```
as.numeric("a") ## NA with a warning
```

```
## Warning: NAs introduced by coercion
```

```
## [1] NA
```

Coercion can be helpful. All the same, it must be used cautiously. Before you move on from this section, if any of this is new, be sure to experiment with different inputs than the ones we tried in the preceding example! Experimenting never hurts, and it can be a powerful way to learn.

Let's turn our attention now to mathematical and logical operators and functions.

1.5 Mathematical Operators and Functions

Several operators can be used for comparison. These will be helpful later, once we get into loops and building our own functions. Equally useful are symbolic logic forms. We start with some basic comparisons and admit to a strange predilection for the number 4:

```
#####################MATH############################
###### Comparisons and logicals
4 > 4
```

```
## [1] FALSE
```

```
4 >= 4
```

```
## [1] TRUE
```

```
4 < 4
```

```
## [1] FALSE
```

```
4 <= 4
```

```
## [1] TRUE
```

```
4 == 4
```

```
## [1] TRUE
```

```
4 != 4
```

```
## [1] FALSE
```

It is sensible now to mention that although the preceding code may be helpful, often numbers differ from one another only slightly—particularly in the programming environment, which relies on the computer representation of floating-point (irrational) numbers. Therefore, we often check that things are close within a tolerance:

```
all.equal(1, 1.00000002, tolerance = .00001)
```

```
## [1] TRUE
```

In symbolic logic, *and* as well as *or* are useful comparisons between two objects. In R, we use & for *and* vs. | for *or*. Complex logic tests can be constructed from these simple structures:

```
TRUE | FALSE
```

```
## [1] TRUE
```

```
FALSE | TRUE
```

```
## [1] TRUE
```

```
TRUE & TRUE
```

```
## [1] TRUE
```

```
TRUE & FALSE
```

```
## [1] FALSE
```

All of the logic tests mentioned so far apply just as well to vectors as they apply to single objects:

```
1:3 >= 3:1
```

```
## [1] FALSE TRUE  TRUE
```

```
c(TRUE, TRUE) | c(TRUE, FALSE)
```

```
## [1] TRUE TRUE
```

```
c(TRUE, TRUE) & c(TRUE, FALSE)
```

```
## [1] TRUE FALSE
```

If you want only a single response, such as for if/else flow control, you can use && or ——, which stop evaluating as soon as they have determined the final result. Work through the following code and output carefully:

```
## for cases where you only want a single response
## such as for if else flow control
## can use && or ||, which stop evaluating after they confirm what it is
## for example
W
```

```
## Error in eval(expr, envir, enclos): object 'W' not found
```

```
TRUE | W
```

```
## Error in eval(expr, envir, enclos): object 'W' not found
```

```
## BUT
TRUE || W
```

```
## [1] TRUE
```

```
W || TRUE
```

```
## Error in eval(expr, envir, enclos): object 'W' not found
```

```
FALSE & W
```

```
## Error in eval(expr, envir, enclos): object 'W' not found
```

```
FALSE && W
```

```
## [1] FALSE
```

Note that the double operators are not, in fact, vectorized. They simply use the first element of any vectors:

```
c(TRUE, TRUE) || c(TRUE, FALSE)
```

```
## [1] TRUE
```

```
c(TRUE, TRUE) && c(TRUE, FALSE)
```

```
## [1] TRUE
```

The any() and all() functions are helpful as well in these contexts for similar reasons:

```
## two additional useful functions are
any(c(TRUE, FALSE, FALSE))
```

```
## [1] TRUE
```

```
all(c(TRUE, FALSE, TRUE))
```

```
## [1] FALSE
```

```
all(c(TRUE, TRUE, TRUE))
```

```
## [1] TRUE
```

We turn our attention now to mathematical, rather than logical, operators. R is powerful mathematically and can perform most mathematical calculations. So although we introduce some functions, we are leaving many out of the mix. For more details, ?Arithmetic can be your friend. It is (as always) important to be aware of the way computers perform mathematical calculations. Being able to code bespoke solutions directly is powerful, yet with the freedom to customize comes a corresponding amount of responsibility. Take a careful look at the following mathematical operations (which can behave differently than expected because of implementation choices):

```
3 + 3
## [1] 6
3 - 3
## [1] 0
3 * 3
## [1] 9
3 / 3
## [1] 1
(-27) ^ (1/3)
## [1] NaN
4 %/% .7
## [1] 5
4 %% .3
## [1] 0.1
```

R also has some common functions that have straightforward names:

```
sqrt(3)
## [1] 1.7
abs(-3)
## [1] 3
exp(1)
## [1] 2.7
log(2.71)
## [1] 1
```

Trigonometric functions also have their part, and ?Trig can bring up a nice list of these. We show cosine's function call cos() for brevity. Note the slight inaccuracy again on the cosine function's output:

```
cos(3.1415) ## cosine
```

```
## [1] -1
```

```
?Trig
```

We close this section and this chapter with a brief selection of matrix operations. Scalar operations use the basic arithmetic operators. To perform matrix multiplication, we use %*%:

```
x2
```

```
##      [,1] [,2]
## [1,]   1    4
## [2,]   2    5
## [3,]   3    6
```

```
x2 * 3
```

```
##      [,1] [,2]
## [1,]    3   12
## [2,]    6   15
## [3,]    9   18
```

```
x2 + 3
```

```
##      [,1] [,2]
## [1,]    4    7
## [2,]    5    8
## [3,]    6    9
```

```
x2 %*% matrix(c(1, 1), 2)
```

```
##      [,1]
## [1,]    5
## [2,]    7
## [3,]    9
```

Matrices have a few other fairly common operations that are helpful in linear algebra. For some of the modeling applications we cover later on, we discuss an appropriate amount of mathematics as needed in the following chapters. Still, this seems a good place to show how the transpose, cross product, and transpose cross product might be coded. We show both the raw code to make the cross product and transpose cross product occur and easier function calls that may be used. This is a relatively common occurrence in R, incidentally. Through packages, quite a few techniques are implemented in fairly clear function calls. Here are the examples:

```
## transpose
t(x2)
```

```
##      [,1] [,2] [,3]
## [1,]   1    2    3
## [2,]   4    5    6
```

```
## cross product
t(x2) %*% x2
```

```
##      [,1] [,2]
## [1,]  14   32
## [2,]  32   77
```

```
## easier cross product
crossprod(x2)
```

```
##      [,1] [,2]
## [1,]  14   32
## [2,]  32   77
```

```
## transpose cross product
x2 %*% t(x2)
```

```
##      [,1] [,2] [,3]
## [1,]  17   22   27
## [2,]  22   29   36
## [3,]  27   36   45
```

```
## easier transpose cross product
tcrossprod(x2)
```

```
##         [,1] [,2] [,3]
## [1,]    17   22   27
## [2,]    22   29   36
## [3,]    27   36   45
```

As you have just seen, it is common in R for someone else to have done the heavy lifting by making a function that outputs the desired outcome. Of course, these friendly programmers' work is subjected to only the underlying constraints of R itself as well as the ability to acquire a free GitHub account. User, beware (at least in some cases)! Thus, it can be helpful to understand the base commands and operators that make R work.

Next, let's focus on understanding implementation nuances as well as quickly getting data in and out of R.

1.6 Summary

We will conclude each chapter with a summary Table 1-2 of any R concepts of major import. These will generally be functions, although some objects will be worth discussing too in the case of this chapter.

Table 1-2. *Chapter 1 summary*

Function	What It Does
factor()	Takes a list of values and assigns a numeric level to them.
c()	Concatenate creates a list of elements.
matrix()	Creates a matrix object.
list()	Creates a list object.
data.frame()	Creates a data frame object.
install.packages()	Installs packages from CRAN by name.
is.data.frame()	Returns a boolean if an object is a data frame.
<-	The assignment operator.
is.integer()	Returns a boolean if an object is an integer.
is.double()	Returns a boolean if an object is a numeric double.
is.vector()	Returns a boolean if an object is a vector.

(continued)

Table 1-2. (*Continued*)

Function	What It Does
is.character()	Returns a boolean if an object is a character.
is.matrix()	Returns a boolean if an object is a matrix.
is.list()	Returns a boolean if an object is a list.
data.table()	Creates a data table object.
is.data.table()	Returns a boolean if an object is a data table.
is.na()	Returns a boolean if an object is NA.
inherits()	Takes two arguments and works when there is not an "is.blank" function already.
as.integer()	Coerces an object to be of type integer.
as.character()	Coerces an object to be of type character.
as.numeric()	Coerces an object to be of type numeric.
as.complex()	Coerces an object to be of type complex number.
as.factor()	Coerces an object to be of type factor.
as.matrix()	Coerces an object to be of type matrix.
as.data.frame()	Coerces an object to be of type data frame.
as.list()	Coerces an object to be of type list.
as.logical()	Coerces an object to be of type boolean.
>	Greater than test returns boolean.
<	Less than test returns boolean.
>=	Greater than or equal to test returns boolean.
<=	Less than or equal to test returns boolean.
==	Equal to test returns boolean.
!=	Not equal to test returns boolean.
all.equal()	Tests if two objects are nearly equal and the third argument is tolerance.
|	Or test returns boolean.
&	And test returns boolean.

CHAPTER 2

Programming Utilities

One of the powerful features of R is the highly skilled, kindly community of enthusiasts, developers, and *package* authors. In particular, to extend the functionality of *base* R, one can find and add *packages* which in turn allow one to use new functions.

As a reminder, in R, functions tend to be actions our code takes to create an *output* or result based on one or more *inputs* (also called *formals*). While we save a discussion for how to code your own functions for another chapter, using functions created and shared in the R community provides highly helpful additions to what R can do.

In particular, we will focus in this chapter on functions for learning more about functions, operating system environment and file management, and data input and output to and from R:

```
options(width = 70, digits = 2)
```

2.1 Installing and Using Packages

Packages are hosted on CRAN [1] which is built into the base R environment (well, technically into a package named utils which is preloaded with base R). There are two functions needed to use a package. One is install.packages(), and this only needs to be run *once per device*. The other is library(), and this is run once per session. We have noted some of our newer-to-R colleagues and students on occasion install packages much more frequently than is required. While there is no particular harm in using the install.packages() function more than necessarily, it can save time to only run it once on each device. In particular, please note that if you already worked through the Chapter 1 content, data.table will already be installed, and thus it is not required to reinstall that package. On the other hand, if you skipped ahead to this chapter (which is of course fine), you will want to install data.table now.

© Matt Wiley and Joshua F. Wiley 2020
M. Wiley and J. F. Wiley, *Advanced R 4 Data Programming and the Cloud*,
https://doi.org/10.1007/978-1-4842-5973-3_2

In the following code, we commented out the installation of our already familiar data.table [9] as well as our new-in-this-chapter haven [33], readxl [27], and writexl [13]. While we run the library() function now, we will discuss each package more in depth when we use it for the first time later inside this chapter. To run commented code, simply do not include the hashtag or pound sign when using that line of code. Please note that R is case sensitive when typing the package and function names.

```
# ch02Packages <- c("data.table", "haven", "readxl", "writexl")
# install.packages(ch02Packages)
library(data.table)
library(haven)
library(readxl)
library(writexl)
```

With that, you now have five packages that add extra features to our journey in this chapter. Notice that it was required to know the *names* of the packages. As a general rule, such names of time- and work-saving packages are learned from social media—#rstats is a common tag on a certain bird-related app—by following certain popular package authors, by reading blogs of various sorts online, or by reading books such as this offering. Many R authors (present company included) and package creators tend to share vignettes and packages via social media.

2.2 Help and Documentation

The R community has many resources to help users. From pre-built functions to whole collections of themed functions in packages, there are many types of support. Both ? and help() are useful ways to access information about an object or function. For more common objects, R has not only extensive documentation about specifics such as input (for functions) but also detailed notes on what those inputs are expected to receive. Furthermore, often detailed examples showcase just what can be done. The Figure 2-1 shows the output of using these two functions with the addition operator:

```
?'+'
help("+")
```

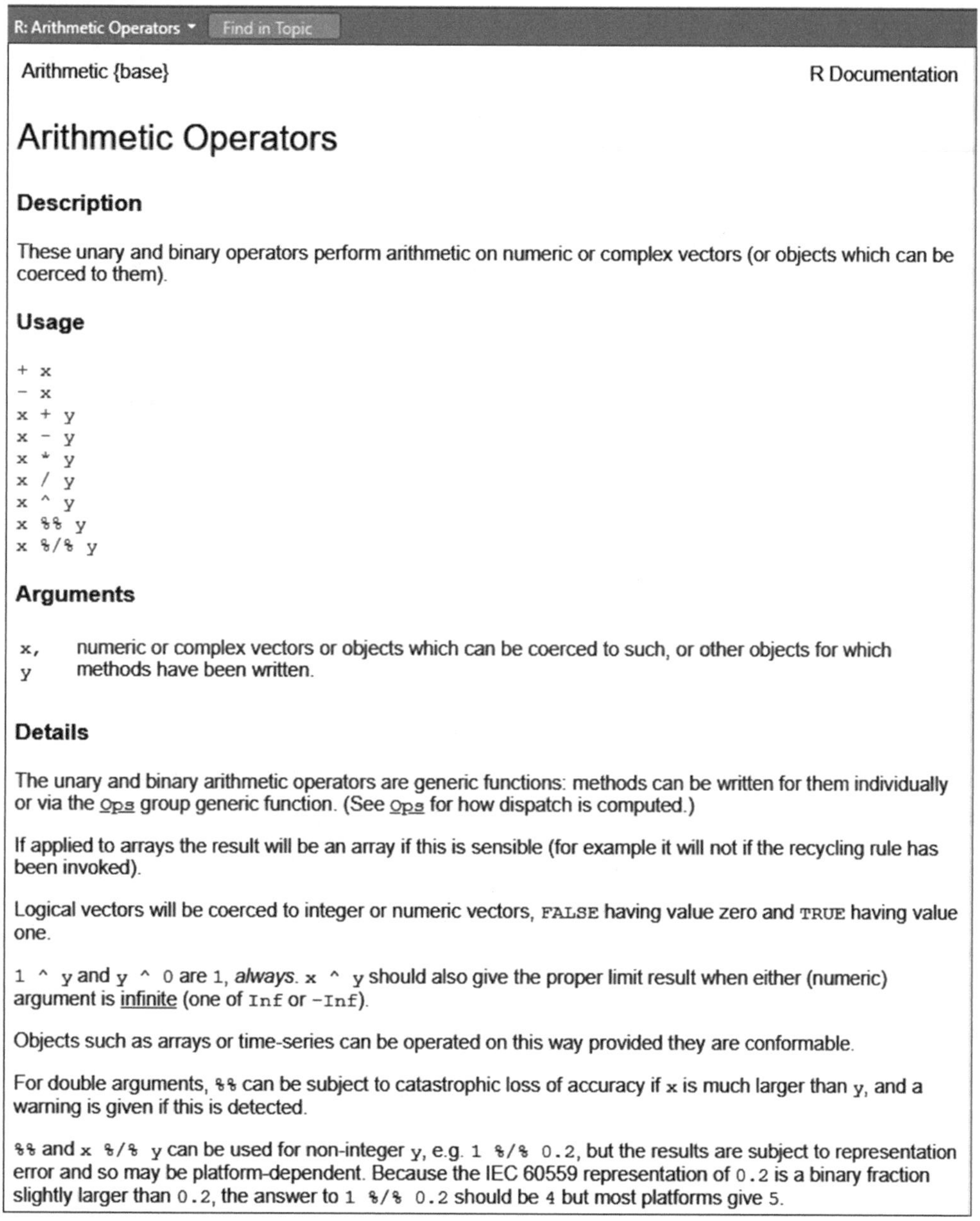

R: Arithmetic Operators ▾   Find in Topic

Arithmetic {base}   R Documentation

Arithmetic Operators

Description

These unary and binary operators perform arithmetic on numeric or complex vectors (or objects which can be coerced to them).

Usage

```
+ x
- x
x + y
x - y
x * y
x / y
x ^ y
x %% y
x %/% y
```

Arguments

x,   numeric or complex vectors or objects which can be coerced to such, or other objects for which
y   methods have been written.

Details

The unary and binary arithmetic operators are generic functions: methods can be written for them individually or via the Ops group generic function. (See Ops for how dispatch is computed.)

If applied to arrays the result will be an array if this is sensible (for example it will not if the recycling rule has been invoked).

Logical vectors will be coerced to integer or numeric vectors, FALSE having value zero and TRUE having value one.

1 ^ y and y ^ 0 are 1, *always*. x ^ y should also give the proper limit result when either (numeric) argument is infinite (one of Inf or -Inf).

Objects such as arrays or time-series can be operated on this way provided they are conformable.

For double arguments, %% can be subject to catastrophic loss of accuracy if x is much larger than y, and a warning is given if this is detected.

%% and x %/% y can be used for non-integer y, e.g. 1 %/% 0.2, but the results are subject to representation error and so may be platform-dependent. Because the IEC 60559 representation of 0.2 is a binary fraction slightly larger than 0.2, the answer to 1 %/% 0.2 should be 4 but most platforms give 5.

Figure 2-1. *An R Manual Help File*

Notice the documentation for arithmetic operators provides specific information about the fact there can be differences in output depending on the platform. It is important to note that not all functions have this fully complete documentation readily available, but for many of the most common functions and packages, an extraordinary level of information is available. Writing code that reproduces the desired results,

29

independent of platform and environment, is not always possible. Later, when we discuss debugging, it is this sort of advanced knowledge of various functions that can be helpful to know.

Of course, this kind of help is more useful when you already know the object or function you want to use and simply need more details. When seeking the ability to do something entirely new, referring to the manuals can help. One such site is `https://cran.r-project.org/manuals.html` maintained by the *R Development Core Team*.

A more common, though less exact, way to gain information is through a suitable search query on your favorite online search engine. Your authors regularly use variations on `R help function-name-here` as a way to locate examples and use cases. Such searches also can include error or warning messages, which can help build deeper knowledge of how functions or their inputs can misbehave.

2.3 System and Files

In a Windows environment, R may be a reasonably effective way to automate file manipulation. IT groups are often willing to install R, and it has the same file permission privileges your username does. From creating files to moving them about and checking dates, R has a variety of functions for getting information from the system and automating file management. Readers who use Unix-based systems usually already have more elegant ways of handling such scenarios from the command line.

One helpful feature of R accessing the system is that it is possible to discover the current date, time, and time zone of the system on which R is being run. This can help detect new files or can be used to put timestamps into files (more on that later). It should be noted that these are, of course, dependent on the system environment being accurate, so caution may be in order before using these in high-stakes projects:

```
Sys.Date()
```

```
## [1] "2020-04-29"
```

```
Sys.time()
```

```
## [1] "2020-04-29 19:13:56 CDT"
```

```
Sys.timezone()
```

```
## [1] "America/Chicago"
```

The help documentation for these commands makes useful suggestions to increase the accuracy of the output, up to potentially millisecond or microsecond precision.

For troubleshooting purposes, the Sys.getenv() function can be very helpful to understand a great many defaults of your R environment and default. While it is safe enough to run the command, due to the machine-specific (and personal) information output, we elect to not show the output here. All the same, it can be interesting to "see under the hood" of your R environment.

As we turn our attention to a variety of file management functions, a word of caution: R has the same privileges you do, and some of these file functions can delete files, can write enough files to fill up a drive, and in general should be handled with a bit of care. That being said, sometimes you need to work with a large collection of files "at scale," and they are all useful, if powerful. These functions share the format file. and have as their first argument a character vector that should be either a filename for the current *working directory* or a path and filename. We start with a text file ch02_text.txt inside two folders data/ch02 under our working directory.

Note   For the following examples, we recommend you have downloaded the full file collection from this book's GitHub site (see details in Chapter 1). If not, then inside your working directory, you will need to create a folder named data; inside that folder, create a subfolder named ch02; and inside that, create a blank text file named ch02_text.txt. In either case, it is not necessarily likely that your output will quite match ours here, since the time of file creation and copy over will change. However, the structure should remain true.

The function file.exists() takes only a character string input, which is a filename or a path and filename. If it is simply a filename, it checks only the working directory. This working directory may be verified by the getwd() function (another function that, while safe to run, we will avoid). If a check is desired for a file in another directory, this may be done by giving a full file path inside the string. This function returns a logical value of either TRUE or FALSE. Depending on user permissions for a particular file, you may not get the expected result for a particular file. In other words, if you do not have permission to access a file (such as on a shared network drive), an existence check may return false, even if the file does exist:

```
# getwd() #safe to run - machine dependent results

file.exists("data/ch02/ch02_text.txt") #meant to return TRUE

## [1] TRUE

file.exists("NOSUCHFILE.FAKE") #meant to return FALSE

## [1] FALSE
```

While the preceding function checks for existence, another function tests for whether we have access to a file. The function file.access() takes similar input, except it also takes a second argument. To test for existence, the second argument is 0. To test for executable permissions, use 1; and to test for writing permission, use 2. If you want to test your ability to read a file, use 4. This function returns an integer vector of 0 to indicate that permissions are given and -1 to indicate permissions are not given. Notice that the default for the function is to test for simple existence. Examples of file.access() are shown here:

```
file.access("data/ch02/ch02_text.txt", mode = 0)

## data/ch02/ch02_text.txt
##                        0

file.access("data/ch02/ch02_text.txt", mode = 1)

## data/ch02/ch02_text.txt
##                       -1

file.access("data/ch02/ch02_text.txt", mode = 2)

## data/ch02/ch02_text.txt
##                        0

file.access("data/ch02/ch02_text.txt", mode = 4)

## data/ch02/ch02_text.txt
##                        0
```

To learn more detailed information about when a file was modified, changed, or accessed, use the `file.info()` function. This function takes in character strings as well. The output gives information about the file size; whether it is a directory; a file permissions integer in read, write, and execute order; the last modified time; the last change time; the last accessed time; and, finally, whether the file is executable:

```
file.info("data/ch02/ch02_text.txt", "NOSUCHFILE.FAKE")
```

```
##                          size  isdir  mode              mtime
## data/ch02/ch02_text.txt    31  FALSE   666  2020-04-25 17:03:35
## NOSUCHFILE.FAKE             NA     NA  <NA>                <NA>
##                                         ctime                atime  exe
## data/ch02/ch02_text.txt 2020-02-15  15:27:15  2020-04-25  17:03:55   no
## NOSUCHFILE.FAKE                            <NA>                <NA> <NA>
```

Using the idea of storing data in a variable from Chapter 1, consider assigning the results of a file to a variable name such as `ch02FileData`. This is in fact a data frame and may be treated as such. Columns can be accessed, values determined, and logical operations performed to understand which files might be above a certain size. In the following code, we read in the system information for two potential files, determine that this is indeed a data frame, and inspect one of the columns more closely, before finally performing some logical tests on that column:

```
ch02FileData <- file.info("data/ch02/ch02_text.txt", "NOSUCHFILE.FAKE")
is.data.frame(ch02FileData)
```

```
## [1] TRUE
```

```
ch02FileData$size
```

```
## [1] 31 NA
```

```
is.numeric(ch02FileData$size)
```

```
## [1] TRUE
```

```
ch02FileData$size > 5
```

```
## [1] TRUE NA
```

In some cases, it can be useful to set various timestamps on certain files. The `sys.setFileTime()` function achieves this, and it takes two inputs. The first is the filename or path and filename. The second input is the time you wish to set on the file. In the following example, we also subtract 20 seconds from the current system time. In this case, it is the `mtime` or *modified time* variable that is changed:

```
file.info("data/ch02/ch02_text.txt")
```

```
##                              size  isdir  mode                    mtime
## data/ch02/ch02_text.txt      31    FALSE  666    2020-04-25    17:03:35
##                                                  ctime           atime   exe
## data/ch02/ch02_text.txt  2020-02-15   15:27:15   2020-04-29   19:13:56    no
```

```
newTime<-Sys.time()-20
newTime
```

```
## [1] "2020-04-29 19:13:36 CDT"
```

```
Sys.setFileTime("data/ch02/ch02_text.txt",newTime)
file.info("data/ch02/ch02_text.txt")
```

```
##                              size  isdir  mode                  mtime
## data/ch02/ch02_text.txt      31    FALSE  666   2020-04-29   19:13:36
##                                                 ctime           atime   exe
## data/ch02/ch02_text.txt  2020-02-15   15:27:15   2020-04-29   19:13:56    no
```

Turning our attention to creation and removal of files, the functions `file.create()` and `file.remove()` do precisely what they say. In addition to performing the indicated action, these functions also return a boolean showing the success or failure of their activity. It is worth mentioning again that removing files should be treated with care. In particular, if you are on a Windows system, try checking the `Recycle Bin` after running the

```
file.create("ch02_created.docx", showWarnings = TRUE)
```

```
## [1] TRUE
```

```
file.remove("ch02_created.docx")
```

```
## [1] TRUE
```

```
file.remove("NOSUCHFILE.FAKE")
```

```
## Warning in file.remove("NOSUCHFILE.FAKE"): cannot remove file
'NOSUCHFILE.FAKE', reason 'No such file
```

```
## [1] FALSE
```

Files may also be copied and renamed. The function file.copy() can be given overwrite permission and could even be set up to copy entire folders and subfolders with the recursive=TRUE option. Furthermore, it has options to copy over mode or file permissions as well as to copy the file date data (or, of course, letting the copy have a new modified date). The following code example shows how that might all work:

```
Sys.time()
```

```
## [1] "2020-04-29 19:13:56 CDT"
```

```
file.copy(
  "data/ch02/ch02_text.txt",
  "ch02_copy.txt",
  overwrite = TRUE,
  recursive = FALSE,
  copy.mode = TRUE,
  copy.date = TRUE
)
```

```
## [1] TRUE
```

```
file.info("ch02_copy.txt")
```

```
##                    size isdir mode                   mtime             ctime
## ch02_copy.txt   31  FALSE 666 2020-04-29 19:13:36 2020-04-29 19:13:56
##                              atime exe
## ch02_copy.txt 2020-04-29 19:13:56   no
```

```
file.rename("ch02_copy.txt", "ch02.txt")
```

```
## [1] TRUE
```

In addition to creating files, creating directories is also called for on occasion. Indeed, creating a new file in a nonexistent directory will not work, so there can be value in leveraging the boolean result of dir.exists() to ensure a directory exists. If not, the function dir.create() can build the necessary folder structure:

```
dir.create("output")
```

```
## Warning in dir.create("output"): 'output' already exists
```

```
dir.exists("output")
```

```
## [1] TRUE
```

2.4 Input

Getting new data into R becomes the next challenge. While in Chapter 1 we saw some ways of building data manually, in general this is not feasible. This chapter shows how to import text files with tab or comma separation as well as some proprietary formats including Excel, SPSS, SAS, and Stata. While we do not cover more exotic data imports in this chapter (e.g., SQL databases or Access or PDF or Word documents), there are packages that handle most types of file imports. For most of these records, the input process is similar, so there is perhaps less of a need to be exhaustive and more of a need to set up sound principles. Be sure to visit the Apress website for this book to download the code and files. In particular, there should be four files named Counties_in_Texas [2] that need to be saved into the two folders data/ch02 under our working directory we used earlier.

We start with a function in data.table, fread(), which reads in .csv or *comma-separated values* (CSV) files. It tends to be very fast—especially for very large files—and actually has several options. While the help files or an Internet search may give you more specific options to your particular use case, the function has four features worth noting. The function takes a header boolean that determines if the first row of data in the file should be treated as column names. It takes a sep or *seperator* argument that determines what the separation character may be (e.g., comma, space, tab). Please take a moment to visit the ?fread help file to see all possible options—it really is a very powerful function. The last two features we left with their default setting; this is worth discussing. Take a look at the data first for a moment, and we will return to those two features:

```
countiesTxCSV <-
  fread("data/ch02/Counties_in_Texas.csv",
        header = TRUE,
        sep = ",")
head(countiesTxCSV)
```

```
##       CountyName FIPSNumber CountyNumber PublicHealthRegion
## 1:     Anderson          1            1                  4
## 2:      Andrews          3            2                  9
## 3:     Angelina          5            3                  5
## 4:      Aransas          7            4                 11
## 5:       Archer          9            5                  2
## 6:    Armstrong         11            6                  1
##       HealthServiceRegion MetropolitanStatisticalArea_MSA
## 1:                   4/5N                              --
## 2:                   9/10                              --
## 3:                   4/5N                              --
## 4:                     11                   Corpus Christi
## 5:                    2/3                    Wichita Falls
## 6:                      1                         Amarillo
##       MetropolitanDivisions MetroArea NCHSUrbanRuralClassification_2006
## 1:                       -- Non-Metro                       Micropolitan
## 2:                       -- Non-Metro                       Micropolitan
## 3:                       -- Non-Metro                       Micropolitan
## 4:                       --     Metro                      Medium Metro
## 5:                       --     Metro                       Small Metro
## 6:                       --     Metro                       Small Metro
##       NCHSUrbanRuralClassification_2013
## 1:                         Micropolitan
## 2:                         Micropolitan
## 3:                         Micropolitan
## 4:                         Medium Metro
## 5:                          Small Metro
## 6:                         Medium Metro
```

In R, functions take *inputs* and use those to create the precise *output(s)* desired. When originally built, the programmer specifies how many arguments, the order those arguments are expected, and a name for those arguments. In the preceding example, fread() relied on the *expected order* of the first argument. In other words, the file location and name are *expected* as the **first** input. However, the next two inputs, where we ensured that the first row of text would become our column names and forced the separation indicator to be a comma, are in fact *not in order*. Because we used the expected variable names for those two arguments however, R properly parsed the function anyway, and the correct things happened. This function would have expected the separator input first. Additionally, the default is `sep = "auto"` which performs some magic that may or may not work to detect the separator character. By changing that to explicitly be a comma, we ensure our file is read correctly. Furthermore, the default is also `header = "auto"`. For many functions that read input, the default is TRUE. Depending on the makeup of the `fread()` automatic process, this may or may not make the correct choice. As of this writing, the automatic process assumes the first line is meant to be column names provided the first row is only character data.

Now, earlier we said there were two more features of this function worth mentioning that we left on the default setting. One of these features is the default choice that `stringsAsFactors = FALSE`. Factors, as mentioned in Chapter 1, are a way of forcing categorical data to have a numerical order. However, on input, the default way to do that ordering would be "in the order the data are read." This may or may not be the correct order to have (e.g., it would be not good for "strongly agree" to code to 1, "neutral" to 2, and "agree" to 3 in most cases). Unlike many `base` R functions for reading CSV input which default to factoring strings, `fread()` defaults to **not factoring**.

The last feature is quite nice, and will become nicer as we explore `data.tables` more. Unlike some functions which read in the data as only a data frame, `fread()` defaults to reading the data in as a `data.table`:

```
is.data.table(countiesTxCSV)
```

```
##  [1] TRUE
```

When working with data, while comma-separated values are common, there are often other formats that need to be read into memory. *Stata* is one such format, such file formats can be identified by the file suffix `.dta`, and the function `read_dta()` comes to us from the `haven` package. Keeping in mind that many more modern versions of R data structures have data frames as their underlying format, with some extras on top, it is worth noting that technically this input is a `tibble`. As mentioned in Chapter 1, we will discuss those more in Chapter 9. For now, it can be treated as any data frame. In this case, while we left it at the default setting of no skipping, we do show explicitly the `skip = 0` option, which can be used to skip any number of top rows (i.e., skip = 2 would skip the first two rows):

```
countiesTxDTA <- read_dta("data/ch02/Counties_in_Texas.dta", skip = 0)
tail(countiesTxDTA)
```

```
## # A tibble: 6 x 10
##    CountyName FIPSNumber CountyNumber PublicHealthReg~ HealthServiceRe~
##    <chr>           <dbl>        <dbl>            <dbl> <chr>
## 1 Wise              497          249                3 2/3
## 2 Wood              499          250                4 4/5N
## 3 Yoakum            501          251                1 1
## 4 Young             503          252                2 2/3
## 5 Zapata            505          253               11 11
## 6 Zavala            507          254                8 8
## # ... with 5 more variables: MetropolitanStatisticalArea_MSA <chr>,
## #   MetropolitanDivisions <chr>, MetroArea <chr>,
## #   NCHSUrbanRuralClassification_2006 <chr>,
## #   NCHSUrbanRuralClassification_2013 <chr>
```

```
is.data.frame(countiesTxDTA)
```

```
## [1] TRUE
```

Similarly, we can also import SPSS files through the read_spss() function. In this case, it seems from our header view that all is well. The only particularly novel feature in the following example is changing the default number of rows to show in header():

```
countiesTxSPSS <- read_spss("data/ch02/Counties_in_Texas.sav", skip = 0)
head(countiesTxSPSS, n = 10)
```

```
## # A tibble: 10 x 10
##     CountyName FIPSNumber CountyNumber PublicHealthReg~
##     <chr>           <dbl>        <dbl>            <dbl>
##  1 Anderson           1            1                4
##  2 Andrews            3            2                9
##  3 Angelina           5            3                5
##  4 Aransas            7            4               11
##  5 Archer             9            5                2
##  6 Armstrong         11            6                1
##  7 Atascosa          13            7                8
##  8 Austin            15            8                6
##  9 Bailey            17            9                1
## 10 Bandera           19           10                8
```

```
## # ... with 6 more variables: HealthServiceRegion <chr>,
## #   MetropolitanStatisticalArea_MSA <chr>,
## #   MetropolitanDivisions <chr>, MetroArea <chr>,
## #   NCHSUrbanRuralClassification_2006 <chr>,
## #   NCHSUrbanRuralClassification_2013 <chr>
```

```
is.data.frame(countiesTxSPSS)
```

```
## [1] TRUE
```

By now, reading data in by file location ought to be fairly mundane. It is worth noting that while our examples have focused on files in subfolders of the working directory, there is no reason why the file location needs to be local. In a network environment with share drives, it is entirely possible to use those remote locations. If the drives are mapped on your local machine to a letter (e.g., an "N:" drive), files in those locations can be accessed in the same way as we do in these examples. A word of caution: Code written with such full file paths tends to *not* be portable to other machines. There are packages and techniques beyond the scope of this book that attempt to solve such issues. For small office environments, standardizing the drive map letter names is one clunky solution to this challenge (provided code is meant to stay in house):

```
countiesTxXl <- read_xlsx("data/ch02/Counties_in_Texas.xlsx")
head(countiesTxXl)
```

```
## # A tibble: 6 x 10
##      CountyName FIPSNumber CountyNumber PublicHealthReg˜ HealthServiceRe˜
##      <chr>        <chr>        <dbl>              <dbl> <chr>
## 1  Anderson       001            1                  4 4/5N
## 2   Andrews       003            2                  9 9/10
## 3  Angelina       005            3                  5 4/5N
## 4   Aransas       007            4                 11 11
## 5    Archer       009            5                  2 2/3
## 6 Armstrong       011            6                  1 1
## # ... with 5 more variables: MetropolitanStatisticalArea_MSA <chr>,
## #   MetropolitanDivisions <chr>, MetroArea <chr>,
## #   NCHSUrbanRuralClassification_2006 <chr>,
## #   NCHSUrbanRuralClassification_2013 <chr>
```

2.5 Output

Much like input, output comes in many forms. Perhaps because of collaboration with other researchers or partners, accommodating one of the other software systems is often needed. Again, R readily outputs to several file types. Often, more than simply converting between file types, the larger interest is in specific console output to certain files. This allows one machine to view the results of an analysis run on another computer. In this section, we demonstrate a couple of outputs of data to SPSS, Stata, and Excel. Then we work with console outputs. We ask you to keep in mind that there are many other ways and types of files to create, and as part of larger examples, we demonstrate several types including various document files and graphics.

Other than the prefix write, these function calls are highly similar to the preceding "Input" section:

```
write_xlsx(countiesTxCSV, "data/ch02/Output1.xlsx") #Excel

write_sav(countiesTxCSV, "data/ch02/Output2.sav") #SPSS

write_dta(countiesTxCSV, "data/ch02/Output3.dta") #Stata
```

The sink() function takes console output and directs it to a file. Thus, the results of an R process may be stored for later observation or saved to a shared drive for perusal by others. The console sends only output to the file, not the input. Calling the function without a file input closes the connection to the file. Look carefully at the difference between the code that follows and the screenshot of Output4.txt shown in Figure 2-2:

```
sink("data/ch02/Output4.txt", append = TRUE, split = TRUE)
x <- 10
xSquared <- x^2

x

## [1] 10

xSquared

## [1] 100

sink()
```

```
Output4.txt - Notepad                              —    □    ×
File Edit Format View Help
[1] 10
[1] 100
```

Figure 2-2. *Screenshot of output text file*

2.6 Summary

In this chapter, we learned to install some external packages to gain additional functions for data input and output. Table 2-1 has a list of the key packages and functions of this chapter.

Table 2-1. *Listing of key functions described in this chapter and summary of what they do*

Function	What It Does
data.table	Nicely reads CSV files.
haven	Reads and writes SPSS and Stata files.
readxl	Reads Excel files.
writexl	Writes Excel files.
?	In front of a function name, opens help manual.
help()	Same as ? above.
Sys.Date()	Shows system date.
Sys.time()	System timestamp.
Sys.timezone()	Local system time zone.
getwd()	Displays working directory.
file.exists()	Returns boolean about named file's existence.
file.access()	Returns 0 for "yes" and -1 for "no" about access rights.
file.info()	Data frame of file size, type, and modification vs. creation vs. access times.

(continued)

Table 2-1. (*continued*)

Function	What It Does
file.create()	Creates a file.
file.remove()	Deletes a file and on Windows may not use Recycle Bin!
file.copy()	Copies a file.
file.rename()	Renames a file without making a new copy.
dir.create()	Creates a new folder or directory.
dir.exists()	Returns boolean about named directory's existence.
fread()	Reads in a CSV file as a data.table.
head()	Shows the first six rows of a data frame.
read_dta()	haven package function reads Stata files.
read_spss()	haven package function reads SPSS files.
tail()	Shows the last six rows of a data frame.
read_xlsx()	readxl package function reads Excel xlsx files.
write_xlsx()	writexl package function writes Excel xlsx files.
write_sav()	haven package function writes SPSS files.
write_dta()	haven package function writes Stata files.
sink()	Writes R console output to file.

CHAPTER 3

Programming Automation

It has been said performing the same action and expecting a different result is a definition of insanity. While a brief search of Oxford's dictionary does not turn up that definition, we have a similar premise: to prevent insanity, leave it to your computer to perform the same action repeatedly!

The goal of this chapter is for you to begin to build automation into your code. Part of the power of code is it cheerfully performs the same action as often as required without stumbling or tiring. The other useful feature of code is the capability to stitch logic into the flow of the programming. Humans, at our best, naturally use such logic. For example, if father's keys are on the hook, he must be home. Otherwise, father must still be out and about. Here's another example: if p is less than alpha, reject the null hypothesis; otherwise, do not reject the null hypothesis.

Because such automation allows an enormous number of repeats, care must be given to efficiency. How long does it take the code to run? Could the code be written differently to make the operations occur faster? In programming, as in life itself perhaps, we often have few perfect answers and more often make trade-offs between choices and consequences. New and more powerful hardware often is cheaper than the human cost required to squeeze out a bit more efficiency; both authors have on occasion solved computation challenges by throwing new hardware at the problem. On the other hand, data sets now come with billions of entries; shaving a millisecond off each calculation saves hundreds of hours of compute time. We do our best to take a balanced approach, demonstrating some of the easier-to-understand constructs first and then presenting faster methods.

As we look at the first of these automation methods, we remind you coding is as much an art as it is a science. The cleaner, more readable code may not be the fastest. The only essential definition of fast is "fast enough." If a particular type of code makes more sense to you and thus makes you more likely to remember and use it, that may be good enough. Of course, if it is not, a bit of research is in order to uncover quicker alternatives.

© Matt Wiley and Joshua F. Wiley 2020
M. Wiley and J. F. Wiley, *Advanced R 4 Data Programming and the Cloud*,
https://doi.org/10.1007/978-1-4842-5973-3_3

3.1 Loops

Loops repeat the code inside them. The concern is to avoid getting into a case of an infinite loop, one that lasts forever. Most loops have a built-in means to attempt a stop (not that those always work as planned). In the next section, we discuss ways to exit a loop manually. Another concern with loops has to do with the fact that they are functions. Although we have used that word already, we are postponing a frank discussion of functions until the next chapter. For now, just keep in mind that anything created inside a function may disappear when that function ends. In the case of loops, this happens when they stop repeating. So, if you want to hold onto results, you need to build the container for those results outside the loop. Again, we go deeper into specifics in the next chapter.

The *for* loop is the first (and perhaps most controversial in R) automator we discuss. In many computer languages, for() is considered rather fast, but not so in R. Nevertheless, this function is easy to understand and use. Human-readable code is not something to eschew needlessly. However, there is usually more than one way to do things, and it would be silly to ignore those. Consider the output of the following line of code, where we use the function print() to print out the values of the variable i:

```
for( i in 1:6){
  print(i)
}

## [1] 1
## [1] 2
## [1] 3
## [1] 4
## [1] 5
## [1] 6
```

What has just occurred is our loop started i with value 1 and then iterated all the way up to 6. This is not especially exciting. However, it shows how the computer is willing to perform some operations one step at a time.

Suppose we wanted to improve the horsepower (hp) of the first six cars in a list. We could add 20 to each of the original values. Using the included mtcars data set, we could go through each hp value one at a time and increase the horsepower:

```
head(mtcars)
```

```
##                    mpg cyl disp  hp drat  wt qsec vs am gear carb
## Mazda RX4           21   6  160 110  3.9 2.6   16  0  1    4    4
## Mazda RX4 Wag       21   6  160 110  3.9 2.9   17  0  1    4    4
## Datsun 710          23   4  108  93  3.8 2.3   19  1  1    4    1
## Hornet 4 Drive      21   6  258 110  3.1 3.2   19  1  0    3    1
## Hornet Sportabout   19   8  360 175  3.1 3.4   17  0  0    3    2
## Valiant             18   6  225 105  2.8 3.5   20  1  0    3    1
```

```
for(i in 1:6 ){
  print(mtcars[i, "hp"] + 20)
}
```

```
## [1] 130
## [1] 130
## [1] 113
## [1] 130
## [1] 195
## [1] 125
```

Of course, this feels a little bit like a toy example, because it is. However, it shows for loops will repeat a tedious task until done.

What is the risk of such a process? Well, it may not be time efficient. Using the function proc.time() in our code to measure different times, we calculate our runtimes by finding the difference between our stop and start times. Additionally, we use the function seq() to create a list of numbers from 1 to 10,000,000 counting up by 1s.

For the for() loop, we initialize an empty variable xCube initialized to an NA value that is of type integer. The loop iterates across the length() of x, in this case, 10 million values. Each value is cubed and stored in our new, empty variable xCube:

```
x <- seq(from = 1,
         to = 10000000,
         by = 1)

forTime <- proc.time() #stores our start time
xCube <- NA_integer_ #creates an empty integer variable
```

```
for(i in 1:length(x)) {
  xCube[i] = x[i] ^ 3 #stores the cube of x at position i in xCube at
  position i
}
```

```
forTime <- proc.time() - forTime #stop time less start time is time taken
forTime #shows total time taken
```

```
##     user  system elapsed
##      3.1     0.2     3.3
```

```
head(x)
```

```
## [1] 1 2 3 4 5 6
```

```
head(xCube)
```

```
## [1] 1 8 27 64 125 216
```

Take a moment to study the preceding example and the output. Most likely, the preceding code took some time to run. If it is taking too much time, simply reduce the number of zeros in your to = argument of your seq() function.

This process took some time because while the cube computation is fast enough, storing data into memory is a slower operation.

This time, though, can be improved. Although we created a variable xCube, it is a null variable. Each iteration of our for loop not only needs to cube x, it also must increase the *length* of the xCube variable by 1. This turns out to not be a trivial task. If we were to create xCube to be already the length we need, performance would significantly improve. We first use the rm() function on xCube, so we have a clean start.

To initialize xCube properly, we use the vector() function, set it to numeric data, and set length to the same length(x) we need. Notice the only difference in this code from the prior example is that xCube starts life as 10 million copies of 0:

```
rm(xCube) #removes the old xCube from memory.
```

```
forTime2<- proc.time()
xCube <- vector(mode = "numeric", length = length(x))
```

```
for (i in 1:length(x) ) {
  xCube[i] = x[i]^3
}

forTime2 <- proc.time()-forTime2
forTime2

##    user  system elapsed
##     1.3     0.0     1.3
```

This improves our time. Of note is even if we had created a dummy xCube that was not quite long enough or if we made one that was too long and needed some null entries deleted off the end, this would remain a faster option. Thus, even when it may not be possible or convenient to know the size of the variable needed before running the loop, it may be beneficial to first create a container mostly large enough to store the data.

Now, for this particular example, a for loop is the wrong way to go. R has powerful and fast underlying code for some of its simplest functions. Notice there is almost no elapsed time at all (of course, there is some, but depending on the operating system, some differences are not noticeable):

```
xCube <- NULL
vTime <- proc.time()
xCube <- x^3
vTime <- proc.time()-vTime
vTime

##    user  system elapsed
##    0.39    0.02    0.41
```

Other types of loops exist. The for loop is best suited to running a process a certain number of times and, when it gets to the end, stopping. The while() loop, on the other hand, is best suited to running a process an uncertain number of times until a stop condition happens. Often, it is possible to gain the same results with different types of loops (or as you have just seen, without a loop at all). All the same, each type of loop tends to have a more natural use under certain conditions. As an exercise on your own, after studying the while loop example we give, see if you can duplicate the previous cubing results.

Perhaps an example near and dear to the hearts of researchers everywhere is simulation. After all, if we can simulate data, that may be the first step to understanding what our real-world results may be. Software-powered simulation often uses a random number generator to create the variation needed. Since we would like our results and your results to match, we explicitly set the same random number starting point via the set.seed(1234) function. While a treatment of random vs. pseudo-random number via computers is beyond the scope of this text, this function ought to get us all on the same page. If you wish to try for truly random inputs, simply remove that single function from your code.

With random numbers discussed, modeling and statistics are baked into R, and the function rnorm() takes up to three inputs. The first controls the number of elements we want to return for our sample. The second and third control the population-level mean and standard deviation from which the sample is randomly selected. Let us a look at what rnorm() provides:

```
set.seed(1234) #this sets the random number start to 1234.

for (i in 1:5 ) {
  x <- rnorm(5,4,2)
  print(x)
  print(paste0("Xbar: ",mean(x)))
  print(paste0("StdDev: ",sd(x)))
}

## [1]  1.59  4.55  6.17 -0.69  4.86
## [1] "Xbar: 3.295292662236"
## [1] "StdDev: 2.78848839771622"
## [1] 5.0 2.9 2.9 2.9 2.2
## [1] "Xbar: 3.17207769924054"
## [1] "StdDev: 1.06732338108581"
## [1] 3.0 2.0 2.4 4.1 5.9
## [1] "Xbar: 3.50884793479832"
## [1] "StdDev: 1.56534479406402"
## [1] 3.8 3.0 2.2 2.3 8.8
## [1] "Xbar: 4.01846923255727"
## [1] "StdDev: 2.76410208454153"
```

```
## [1] 4.3 3.0 3.1 4.9 2.6
## [1] "Xbar: 3.5874894580705"
## [1] "StdDev: 0.965711540327254"
```

What we find are five samples randomly pulled from the same population with a mean() of 4 and a standard deviation or sd() of 2. However, notice each sample has different means. The means are all close to 4 of course; it would not be a normal deviation otherwise. Indeed, we might expect most individual numbers to have an average between 2 and 6 and almost all to be between 0 and 8. Since we are looking at the mean of five numbers, central limit theorem suggests it would be fairly strange to be too far away from the mean.

Suppose we wanted to find an example of five elements randomly drawn from the population which has a normal distribution of mean 4 and standard deviation of 2. How many samples would we need to randomly pull before we found the set that had a mean of at least 8.1?

The answer is a while() loop. Unlike a for loop which runs for a *fixed* number of iterations, a while loop keeps running until some condition is met. In our case, we want to keep pulling a random set of five numbers from our distribution until we get a subset with a mean of at least 8.1 (a value chosen arbitrarily simply because it is quite a few standard deviations above the expected).

We again fix ourselves to a specific random number sequence. To ensure our loop runs at least once, we must ensure that mean(y) < 8.1 returns TRUE. This is achieved by setting y <- 8 as a starting value. Additionally, just for learning purposes, we add in a zeroed-out counter value i that counts how many times our loop runs. Lastly, we confirm our randomly pulled sample does indeed have a mean that is quite large given our population's distribution:

```
set.seed(1234) #allows duplication of results

y <- 8 #mean(y) < 8.1 returns TRUE
i <- 0 #initialise counter and set to zero

while (mean(y) < 8.1) {
  i <- i + 1 #increase counter by 1 each time
  y <- rnorm(5, mean = 4.0, sd = 2) #store random sample from population
}
```

```
print(y) #observe sample
```

```
## [1] 10.3   9.0   7.7   6.7   8.3
```

```
mean(y) #confirm mean(y) >= 8.1
```

```
## [1] 8.4
```

Notice y is a rather strange sample, compared to what we know the population to be in general. This is, of course, an argument for larger sample sizes, but it also gives us data that both fits our facts and tests our assumptions. Counterexamples and odd data can be quite useful to stress test code or models. This is only one example of a while loop. As you can see, the argument is where the condition is set. The loop will keep running until the condition turns FALSE. Indeed, this loop had to run several times; our counter is well over 1 million loops!

The last loop we consider is repeat() which can be different from while and for loops. In particular, repeat, unlike while or for, does not take any arguments that would ever have a chance to stop it. This function keeps repeating over and over. Of course, there has to be a way to stop this. Otherwise, we are in trouble. We discuss such techniques in the next section, "Flow Control." However, first, here are some observations about loops.

Loops are meant to perform a task as many times as required. In R, we control these tasks in three ways. If the task should be done to a certain count, a for loop is likely best. A fixed iterator controls the start and stop of the for loop. If the task is one we want to be done until a certain condition occurs, a while loop may be best. In particular, a while loop tests the condition that determines whether the loop runs, before the loop runs. This is critical! If the condition is already met, the while loop never runs. The last loop we discuss, repeat(), is different. It simply runs. Now, we have to stop it somehow, and we manually stop a repeat in the next section. The point is repeat never checks whether it needs to run; it just runs. So if you want to run a process at least once, repeat may be the function you seek.

We hold off on giving an example of a repeat for now, mostly to avoid infinite loops. While we will show an example soon, we must first learn a way to control the flow of our code.

3.2 Flow Control

Code flow control works by performing tests to decide whether one action or another should be taken or by modifying the behavior of the loops you saw in the prior section. We present each of the commands in turn, and, at the end of this section, we return to the repeat loop.

If/else statements are a standard part of logic, and they make up a standard part of programming as well. The if/else control tests a logical condition which, when TRUE, the if portion runs. Otherwise, when the logical condition is FALSE, the else portion runs. Sometimes, there is not an alternative bit of code that needs to be run; in that case, the else is simply not used.

Going back to our sample of values for which we forced a sample mean greater than or equal to 8.1, we can use if/else to determine what the title of the plot says (in other words, we can use code logic to change the output of our graphs).

The following code generates Figure 3-1 using flow control. The logical operation mean(y) > 8.1 resolves to TRUE. This in turn runs the function boxplot() and *skips* the plot() function. The boxplot() takes the entire data set as the first input and takes a main title for the second, and we set the horizontal value to TRUE away from the default of FALSE:

```
if( mean(y) > 8.1 ) {

  boxplot(y,
          main = "This sample has an atypically large mean.",
          horizontal = TRUE)

} else{

  plot(y,
       main = "This sample mean is not so very large.")
}
```

This sample has an atypically large mean.

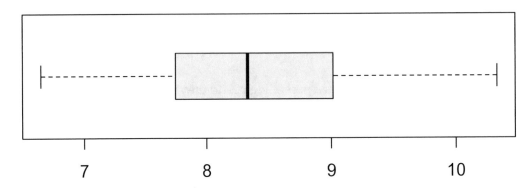

Figure 3-1. Box and whisker plot showing our atypical sample

Our next flow control function is next() which can be used to move on to the next iteration through a loop. In other words, next() will skip past and not use the code that follows it. However, it does not end or stop the looping process. Recall our while() loop from earlier. Compare and contrast that earlier version of the code to the code that follows. In particular, note that the print(y) has been moved to inside the loop. And yet, despite the loop running over 1.2 million times, the value of the y sample is only output to the console once:

```
set.seed(1234) #allows duplication of results

y <- 8 #mean(y) < 8.1 returns TRUE

while (mean(y) < 8.1) {
  y <- rnorm(5, mean = 4.0, sd = 2) #store random sample from population

  if ( mean(y) < 8.1 ) { next() }
  print(y)
}

## [1]  10.3  9.0  7.7  6.7  8.3
```

The location of next is significant; we could order those three lines of code inside the while function in six ways. At least one way creates an infinite loop, and at least one way has no noticeable influence whatsoever. Of course, as you just saw, there is also a way that makes this loop behave similarly to some prior code. An important feature of next() is we can now choose (when combined with if) whether we want to complete every single loop or only some of the loops.

This leads us to a discussion of break, which not only stops one cycle of a loop—it *breaks* us out of the loop altogether. When paired with the repeat() function, it is possible to create a loop that will keep running until it reaches a certain condition.

While this code has the same net result as the two prior versions, notice that our logic structure ends up being just a bit different. When the break criteria is reached in the if statement, our repeat loop—which has no native iteration control—will be forcibly broken. Note this break does *not* prevent the print(z) from working. This only breaks the flow of the code out of the loop—it does not quit the entire run:

```
set.seed(1234) #allows duplication of results

z <- NULL

repeat{
  z <- rnorm(5, mean = 4.0, sd = 2) #store random sample from population
  if (mean(z) >= 8.1) { break }
}

print(z)

## [1]  10.3  9.0 7.7  6.7  8.3
```

3.3 apply Family of Functions

Our last set of functions for this chapter are the *apply functions. Generally speaking, these functions take two primary types of input. One is an input that has several elements, and the second is a function applied to the elements in turn. The various flavors handle different use cases, and we also introduce some error checking where possible.

We start with lapply, which takes an input and applies the function given to each element. As its prefix suggests, this function returns a list with the results of the application of the entered function. In this case, we apply the sqrt() or square root function to each element of our list():

```
xL <- 1:10
lapply(xL, sqrt)
```

```
## [[1]]
## [1] 1
##
## [[2]]
## [1] 1.4
##
## [[3]]
## [1] 1.7
##
## [[4]]
## [1] 2
##
## [[5]]
## [1] 2.2
##
## [[6]]
## [1] 2.4
##
## [[7]]
## [1] 2.6
##
## [[8]]
## [1] 2.8
##
## [[9]]
## [1] 3
##
## [[10]]
## [1] 3.2
```

As you can see, this is fairly messy. To simplify, we use almost the same function call, but with the s prefix (for simple). In particular, the output of sapply() is generally either a vector or a matrix object. In this case, the returned object is a vector:

```
sapply(xL, sqrt)
```

```
## [1]  1.0  1.4  1.7  2.0  2.2  2.4  2.6  2.8  3.0  3.2
```

Although both of these are helpful enough in their own right, they depend on correct input. Of course, code in general depends on such correct input (the phrase "garbage in, garbage out" comes to mind). Nevertheless, we can signal the type of results expected with the vapply() function. This function takes a third input, which is set to the type of output we are expecting. What happens when the function does not return the correct type of value? Here we run two almost similar lines of code. The first line tells vapply() to expect boolean outputs because by default NA is a logical. Our second attempt expects single values of type double() as outputs. The function double(length = 1) is expecting each object of type double to be a single double element. As this is precisely what our apply function is creating as output, the next example works:

```
vapply(xL, sqrt, FUN.VALUE = NA) #this has an error
```

```
## Error in vapply(xL, sqrt, FUN.VALUE = NA): values must be type 'logical',
## but FUN(X[[1]]) result is type 'double'
```

```
vapply(xL, sqrt, FUN.VALUE = double(1))
```

```
##  [1] 1.0 1.4 1.7 2.0 2.2 2.4 2.6 2.8 3.0 3.2
```

We run one last similar line of code, this time telling R to expect each result to be a vector of type double and length 2. Instead, it gets a vector of length 1, which is the correct type, yet wrong length:

```
vapply(xL, sqrt, FUN.VALUE = double(2)) #this has an error
```

```
## Error in vapply(xL, sqrt, FUN.VALUE = double(2)): values must be length 2,
##  but FUN(X[[1]]) result is length 1
```

The ability of vapply to provide some sort of native error check can be rather useful. Now that the basics of the apply family of functions are understood with these toy examples, let us consider a more interesting data set. Notice the data set iris has both integers and characters. Now suppose we are interested in summary statistics information for each element in iris. We could certainly apply the summary() function to each element, and R would cheerfully do so and not even throw an error at us!

```
lapply(iris, summary)

## $Sepal.Length
##    Min. 1st Qu. Median    Mean 3rd Qu.    Max.
##    4.3     5.1    5.8     5.8     6.4     7.9
##
## $Sepal.Width
##    Min. 1st Qu. Median    Mean 3rd Qu.    Max.
##    2.0     2.8    3.0     3.1     3.3     4.4
##
## $Petal.Length
##    Min. 1st Qu. Median    Mean 3rd Qu.    Max.
##    1.0     1.6    4.3     3.8     5.1     6.9
##
## $Petal.Width
##    Min. 1st Qu. Median    Mean 3rd Qu.    Max.
##    0.1     0.3    1.3     1.2     1.8     2.5
##
## $Species
##     setosa versicolor virginica
##         50         50        50
```

While this provided the type of five-figure summary we might expect for several rows, the last row is not like the rest. And since the `iris` data set is short enough that a visual inspection is enough to know the last row is not like the others, it is possible a longer data set might not be as easy to inspect visually. Thus, the error checking feature of `vapply()` can be quite helpful. The output can be more explicitly described than simply "data type." Instead, we can specify the format by spelling out the names and types of data expected. The first line of code uses the entire iris data frame, while the second example only uses the first four columns (and thus works):

```
vapply(iris, summary, c(Min. = 0, '1st Qu.' = 0, Median = 0,
                        Mean = 0, '3rd Qu.' = 0, Max. = 0))

## Error in vapply(iris, summary, c(Min. = 0, '1st Qu.' = 0, Median = 0, :
values must be length 6,
##   but FUN(X[[5]]) result is length 3
```

```
vapply(iris[1:4], summary, c(Min. = 0, '1st Qu.' = 0, Median = 0,
                             Mean = 0, '3rd Qu.' = 0, Max. = 0))
```

```
##          Sepal.Length Sepal.Width Petal.Length Petal.Width
## Min.              4.3         2.0          1.0         0.1
## 1st Qu.           5.1         2.8          1.6         0.3
## Median            5.8         3.0          4.3         1.3
## Mean              5.8         3.1          3.8         1.2
## 3rd Qu.           6.4         3.3          5.1         1.8
## Max.              7.9         4.4          6.9         2.5
```

Another useful data set built into R is the mtcars data frame from a 1974 magazine. We are interested in the miles per gallon and the number of cylinders. Utilizing this data set is a useful way to understand various functions in R, as it is readily accessible and does not need to be imported. Here, we take a look at the first six rows to see and better understand our raw data. Additionally, we see there are three types of vehicle cylinders, those with four, six, and eight cylinders:

```
head(mtcars)
```

```
##                    mpg cyl disp  hp drat  wt qsec vs am gear carb
## Mazda RX4           21   6  160 110  3.9 2.6   16  0  1    4    4
## Mazda RX4 Wag       21   6  160 110  3.9 2.9   17  0  1    4    4
## Datsun 710          23   4  108  93  3.8 2.3   19  1  1    4    1
## Hornet 4 Drive      21   6  258 110  3.1 3.2   19  1  0    3    1
## Hornet Sportabout   19   8  360 175  3.1 3.4   17  0  0    3    2
## Valiant             18   6  225 105  2.8 3.5   20  1  0    3    1
```

```
unique(mtcars$cyl)
```

```
## [1] 6 4 8
```

To more closely understand how our next apply function works, we first use the plot() function to visualize cyl on an x-axis and mpg on a y-axis as seen in Figure 3-2.

```
plot(mtcars$cyl, mtcars$mpg)
```

To use `tapply()`, we specify the data the function is applied to, the index of those data, and the function to be applied. In this case, we wish to better understand the arithmetic mean of each of the three types of cars. Keeping in mind Figure 3-2, the results should be roughly expected as to which types of calendars get the best mileage:

```
tapply(
  X = mtcars$mpg,
  INDEX = mtcars$cyl,
  FUN = mean)
```

```
##   4   6   8
## 27  20  15
```

When our data structure is a data frame, matrix, or array, the `apply()` function is often a good choice. Because these tend to have more dimensions, this function requires more information about which data the function is going to use as input. In the second formal argument, named `MARGIN`, an input value of 1 specifies *rows*, while a value of 2 specifies *columns*. For a data frame such as `mtcars`, most computations only make sense by column, and thus `MARGIN = 2` is used. In this case, rather than mean, we use the standard deviation function `sd()`:

```
apply(mtcars, MARGIN = 2, FUN = sd)
```

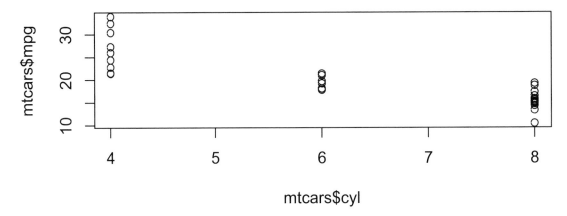

Figure 3-2. *Visualizing the data of cyl vs. mpg*

```
##     mpg    cyl    disp     hp   drat     wt   qsec     vs     am   gear
##    6.03   1.79  123.94  68.56   0.53   0.98   1.79   0.50   0.50   0.74
##    carb
##    1.62
```

When needing to apply the same function to more than one set of data, the multivariate apply mapply() is required. Unlike the prior examples, this one takes the function as our first variable and from there takes as many data inputs as provided. In this case, rather than double up on function calls, it is possible to perform the entire computation in a single pass.

First, consider a single correlation test comparing mpg to cyl using the cor.test() function:

```
cor.test(mtcars$mpg, mtcars$cyl)
```

```
##
##        Pearson's product-moment correlation
##
## data: mtcars$mpg and mtcars$cyl
## t = -9, df = 30, p-value = 6e-10
## alternative hypothesis: true correlation is not equal to 0
## 95 percent confidence interval:
##   -0.93 -0.72
## sample estimates:
##    cor
## -0.85
```

The results indicate fairly strong evidence for negative correlation. Now suppose, in addition to this comparison, we also wish to test the relationship between hp and disp. Rather than perform both tests one at a time, the mapply() function will handle this computation. Notice we are testing the same two pairs. In other words, we now have Table 3-1. Our function mapply() computes cor.test() on the first column and then on the second column of the table. Additional arguments can be fed to cor.test() via the MoreArgs = variable which takes a list. In this case, while we did explicitly show these inputs, we left them on their default values:

```
mapply(cor.test,
       mtcars[, c("mpg", "hp")],
       mtcars[, c("cyl", "disp")],
```

Table 3-1. *mapply() grid*

First	Second
mpg (1,1)	hp (1,2)
cyl (2,1)	disp (2,2)

```
MoreArgs = list(method = "pearson", alternative = "two.sided"))
```

```
##                 mpg
## statistic       -8.9
## parameter       30
## p.value         6.1e-10
## estimate        -0.85
## null.value      0
## alternative     "two.sided"
## method          "Pearson's product-moment correlation"
## data.name       "dots[[1L]][[1L]] and dots[[2L]][[1L]]"
## conf.int        Numeric,2
##                 hp
## statistic       7.1
## parameter       30
## p.value         7.1e-08
## estimate        0.79
## null.value      0
## alternative     "two.sided"
## method          "Pearson's product-moment correlation"
## data.name       "dots[[1L]][[2L]] and dots[[2L]][[2L]]"
## conf.int        Numeric,2
```

One part that is admittedly a touch confusing about the results is the data.name row. However, a review of Table 3-1 should allow that to make more sense. The feature to notice is mapply applied the given function to the first entry in each "row" of input.

3.4 Summary

You have seen several powerful tools for automation, from loops of various sorts that control what happens repeatedly (and, indeed, how often) to more specialized tools streamlined to cope with the usual suspects in data sets that we are likely to encounter. These techniques are helpful and are often used. However, they are not enough. What if new techniques are developed? How does a function like mean() work? What if we need to create our own functions? In the next chapter, we explore functions both in terms of their theoretical framework and their practical applications. The combination of custom functions and loops is the bread and butter of coercing computers to do the busy work for us. Until then, take a close look at the functions learned in this chapter in Table 3-2, and consider what might live "under the hood."

Table 3-2. *Listing of key functions described in this chapter and summary of what they do*

Function	What It Does
for()	Invokes a for loop and as input requires an index along with start/stop points.
print()	Prints argument to console.
head()	By default shows first six rows of a data frame.
seq()	Generates a number sequence from a starting point to an ending point by a fixed "hop."
proc.time()	Provides time information to evaluate code efficiency.
length()	Measures the number of entries in a vector.
rm()	Removes the specified object from environment memory.
rnorm()	Generates a random sample with specified number of elements, from a population with specified mean and standard deviation.
set.seed()	Locks the random number generator to a specified starting point—allows different computers to get the same "random" results.
paste0()	Merges two or more strings together with no space between them.
mean()	Computes the arithmetic mean.

(*continued*)

Table 3-2. (*continued*)

Function	What It Does
sd()	Computes the standard deviation.
while()	Invokes a while loop and as input requires a stop criteria boolean.
repeat()	Invokes a repeat loop that takes no input. Requires an internal break condition to avoid infinite looping.
if()else	Evaluates criteria input. When TRUE runs the if option and otherwise runs the else.
boxplot()	Draws a boxplot from the input data.
plot()	Draws an x/y-axis graph from input x values and y values.
next()	Used inside a loop to skip any following lines of code—usually used in conjunction with an if.
break()	Used inside a loop to stop the loop entirely and break out of the loop process—usually used in conjunction with an if.
lapply()	To the given list, applies the given function.
sqrt()	Returns the square root of the given value.
sapply()	To the given list, applies the given function and returns a simplified object.
vapply()	To the given list, applies the given function and checks for value matching.
summary()	Computes a six-element summary of input numbers (min/max plus quartiles and mean).
unique()	Reduces an input object to unduplicated elements.
tapply()	Groups the given list of data by the unique values of the given index and applies the given function.
apply()	To the given data frame, by either row (1) or column (2), applies the given function.
cor.test()	Performs a t test on correlation.
mapply()	Applies the given function to the given data sets by column.

CHAPTER 4

Writing Functions

Writing your own functions in R enables you to combine a set of R commands into a function that is easy to call and can be generalized. Functions are foundational to R. To become a more advanced user or developer of R, a good understanding of what functions are and how to write them is crucial. Broadly speaking, a function takes one or more inputs and processes them to produce and return output.

Not every programming task should be converted to a function. However, whenever you find yourself copying and pasting a particular line of your code for the third time, you should likely write a function. Another "no question about it" time to write a function is when your code is over 100 lines or so. Long chunks of code can become almost impossible to read through and understand. Instead, we write functions with good, descriptive names that make our code more readable. Then, on a separate pass, we write code using the functions. The beauty of such a system is that it becomes easier to write focused code inside each function that solves one particular part of your challenge. Another benefit of this approach is that, should greater efficiency ever be required, it is possible to determine which functions are costing you the most processing power or take the longest time to complete. Then research can be done on how to make that specific stage more efficient.

In this chapter, we use the Hmisc [10] R package:

```
#install.packages("Hmisc")
library(Hmisc)
options(width = 70, digits = 2)
```

© Matt Wiley and Joshua F. Wiley 2020
M. Wiley and J. F. Wiley, *Advanced R 4 Data Programming and the Cloud*,
https://doi.org/10.1007/978-1-4842-5973-3_4

4.1 Components of a Function

With some exceptions, most functions in R have three components:

- Formals: The arguments, or inputs, to the function

- Body: The commands that process the input

- Environment: The location or context of a function, which determines where it looks for variables

Each can be examined using dedicated functions. First, we write an example function:

```
f <- function(x, y = 5) {
  x + y
}
```

The function accepts two arguments: x and y. The first argument, x, has no default, but the second argument, y, defaults to a value of 5. Although these two variables are easily seen (in part because we just wrote the function), the *formals*, or arguments, of the function can be examined using the formals() and args() functions:

```
formals(f)
```

```
## $x
##
##
## $y
## [1] 5
```

```
args(f)
```

```
## function (x, y = 5)
## NULL
```

Then, to see the actual R code or the commands used to process the formals, we use the body() function:

```
body(f)
```

```
## {
##     x + y
## }
```

The commands are always enclosed in opening and closing brackets. In this case, the R code is simply x + y. However, some functions have hundreds of lines. The last key part of a function is its environment. The environment of a function determines where it looks for variables or objects, which can include both data and functions. To see the environment of a function, we use the environment() function:

```
environment(f)
```

```
## <environment: R_GlobalEnv>
```

In this case, the function is in the global environment, where we created the function. For other functions, this would vary. For example, if we look at the environment for the plot() function, it is the namespace for the graphics [16] package which is included by default in R:

```
environment(plot)
```

```
## <environment: namespace:base>
```

This provides a common language for discussing R functions. In the remainder of the chapter, we delve deeper into writing functions and the special code and tools available for use with functions.

4.2 Scoping

In R, *scoping* is what determines where to look for a particular variable. Consider, for example, when we type plot. How does R translate that code? Where does R look it up? Scope is R's answer for the language idea of *context*. If one says, "I bought a new bat," you would likely suppose a cricket bat rather than a flying mammal. For R, context comes from the scope and environment. Thus, different environments can lead to different results.

Most aspects of writing R functions are no different than using R interactively. However, one difference is functions have their own environment. Further, functions often are located *in* various environments. For instance, when using R interactively, almost all commands are executed from the global environment. In contrast, many functions are written as part of R packages, in which case the function's environment is defined by the R package. Therefore, before jumping into writing functions, it is helpful to understand scoping. Consider these two examples:

plot

```
## function (x, y, ...)
## UseMethod("plot")
## <bytecode: 0x000002ac822a6798>
## <environment: namespace:base>
```

plot <- 5
plot

```
## [1] 5
```

In the first instance, R finds plot in the graphics package. In the second instance, R finds plot in the global environment. Assigning 5 to the variable, plot, does not overwrite the plot() function. Instead, it creates another R variable object with the same name as the function. After creating the new variable, when we type plot at the console, R returns the numeric value rather than the function because the assignment, plot <- 5, occurs in the global environment, which R searches before it checks the environments of different packages. The search() function returns the environments in the order that R searches them. Your environment may be different, although there are likely some similarities:

```
search()
```

```
##  [1] ".GlobalEnv"         "package:Hmisc"     "package:ggplot2"
##  [4] "package:Formula"    "package:survival"  "package:lattice"
##  [7] "package:writexl"    "package:readxl"    "package:haven"
## [10] "package:data.table" "package:knitr"     "tools:rstudio"
## [13] "package:stats"      "package:graphics"  "package:grDevices"
## [16] "package:utils"      "package:datasets"  "package:methods"
## [19] "Autoloads"          "package:base"
```

From the output, we see R first looks in the global environment .GlobalEnv, then in the package Hmisc, and so on until it reaches the base package. We can see the current environment by again using the environment() function. R always begins looking in the current or local environment. From there, it progresses to the parent environment. We can find the parent environment for a given environment by using the parent.env() function:

```
environment()
```

```
## <environment: R_GlobalEnv>
```

```
parent.env(.GlobalEnv)
```

```
## <environment: package:Hmisc>
## attr(,"name")
## [1] "package:Hmisc"
## attr(,"path")
## [1] "C:/.../R/win-library/4.0/Hmisc"
```

Coming back to functions, each function has its local environment in addition to the function itself being in an environment. The following code shows the local function environment and its parent environment. We can roughly classify variables in functions into one of three types:

- Formal variables

- Local variables defined within the function

- Free or other variables that are neither formals nor local variables

Each is demonstrated in turn with the following code:

```
a <- "free variable"
f <- function(x = "formal variable") {
  y <- "local variable"

  e <- environment()
  print(e)
  print(parent.env(e))

  print(a)
  print(x)
  print(y)
}

f()
```

```
## <environment: 0x000002ac9061f258>
## <environment: R_GlobalEnv>
```

```
## [1] "free variable"
## [1] "formal variable"
## [1] "local variable"
```

The variable x is a formal defined as its default value, y is a local variable defined in the body of the function, and a is a variable defined in the function's parent environment. Although it is possible to rely on objects in the search path for a function rather than identified in the formals or as a local variable, it is not a wise idea. Coding to depend on a function's parent environment or the search path can lead to particularly tricky bugs and unexpected behavior. This creates chaos for users or yourself later, when something in the environment seemingly unrelated to the function is changed, and even though it appears the function's code and inputs have not changed, suddenly the output is different.

Although the examples so far have been relatively straightforward, scoping becomes trickier when using nested function calls. In the following code, it may be harder to predict the results of each piece of evaluated code. The next examples set up two functions and then use the functions with three different variables in the global environment:

```
f1 <- function(y = "f1 var") {
  x <- y
  a1 <- f2(x)
  rm(x)
  a2 <- f2(x)
}
f2 <- function(x) {
  if (nchar(x) < 10) {
    x <- "f2 local var"
  }
  print(x)
  return(x)
}

x <- "global var"
f1()

## [1] "f2 local var"
```

```
## [1] "global var"
```

```
x <- "g var"
f1()
```

```
## [1] "f2 local var"
## [1] "f2 local var"
```

```
rm(x)
f1()
```

```
## [1] "f2 local var"
```

```
## Error in f2(x): object 'x' not found
```

It is worth experimenting with scoping until it makes sense, because it is necessary for ensuring the correct object is found. This can be a particular challenge when writing and developing a package or when using functions that appear in multiple packages.

4.3 Functions for Functions

In addition to the usual R code and functions you may use within a function you write, some special functions exist. These are only, or primarily, used within other functions. Even if you are an experienced R user, these may be unfamiliar if you have not previously written functions.

The match.arg() function is useful for performing fuzzy matching of arguments. This also has the benefit that if an argument does not match one of the valid options (ergo is invalid), it throws an error. The following code shows two similar functions, except one uses the match.arg() function. The examples demonstrate fuzzy matching and what happens when an invalid argument is used:

```
f1 <- function(type = c("first", "second")) {
  type
}
f2 <- function(type = c("first", "second")) {
  type <- match.arg(type)
  type
}
f1("fi")
```

```
## [1] "fi"
```

```
f2("fi")
```

```
## [1] "first"
```

```
f1("test")
```

```
## [1] "test"
```

```
f2("test")
```

```
## Error in match.arg(type): 'arg' should be one of "first", "second"
```

Argument matching is also useful for ensuring that when a text string is passed, only a valid option is used, and if a valid option is not used, an informative error is thrown. The following code expands on the function we built without match.arg() to calculate a mean if type = "first" and a standard deviation if type = "second". However, when an invalid option is passed to the type argument, we get the relatively cryptic error message that x is not found:

```
f1b <- function(type = c("first", "second")) {
  if (type == "first") {
    x <- mean(1:5)
  } else if (type == "second") {
    x <- sd(1:5)
  }
  return(x)
}
```

```
f1b("test")
```

```
## Error in f1b("test"): object 'x' not found
```

Another function specific to function arguments is missing(). It returns a logical value indicating whether a specific argument is missing from the function call. This is helpful when writing functions that may be used in different ways. The following is an example of a function that calculates *Cohen's d* effect size, which for a single group is defined as the mean of a variable divided by the standard deviation. Cohen's d can also be calculated for repeated measures, such as a group measured before and after an intervention. This is done by calculating the difference of the two variables and then proceeding as before. We use the `missing()` function to determine whether the user passes a single variable, x, or two variables, x and y, so our function can elegantly handle calculating Cohen's d for both one sample and repeated-measures data:

```
cohend <- function(x, y) {
  if (!missing(y)) {
    x <- y - x
}

  mean(x) / sd(x)
}

cohend(x = c(0.61, 0.99, 1.47, 1.52, 0.45, 3.34, 1.05, -1.47, 1.3, 0.33),
       y = c(-0.69, 1.6, 0.44, 1, 0.88, 1.17, 2.4, 1.21, 0.87, 2.15))

## [1] 0.095

cohend(x = c(0.61, 0.99, 1.47, 1.52, 0.45, 3.34, 1.05, -1.47, 1.3, 0.33))

## [1] 0.8
```

Also related to determining characteristics of the function call is the function `match.call()`. Whereas `missing()` determines whether a specific argument is missing and `match.arg()` determines whether an argument matches one of the valid options, `match.call()` captures the entire function call. This might be easier to demonstrate than to explain. The following little function calculates the coefficient of variation, the sample standard deviation divided by the sample mean. It also captures and returns the function call by using `match.call()`:

```
cv <- function(x, na.rm = FALSE) {
  fcall <- match.call()

  est <- sd(x, na.rm = na.rm) / mean(x, na.rm = na.rm)

  return(list(CV = est, Call = fcall))
}

cv(1:5)

## $CV
## [1] 0.53
```

```
##
## $Call
## cv(x = 1:5)

cv(1:8, na.rm = TRUE)

## $CV
## [1] 0.54
##
## $Call
## cv(x = 1:8, na.rm = TRUE)
```

In the output from those two examples, `match.call()` captures exactly the call to the function, though it adds explicit argument names. This can sometimes be useful for keeping a record of exactly what the call was that created particular output. Perhaps the most common place where this is used is in the output from regression models in R. For example, the following code shows a linear model in which the output echoes the function call, which is done using `match.call()`:

```
lm(formula = mpg~hp, data = mtcars)

##
## Call:
## lm(formula = mpg ~ hp, data = mtcars)
##
## Coefficients:
## (Intercept)              hp
##      30.0989        -0.0682
```

The `return()` function is typically used at the end of functions to return a specific object, as you have already seen in some of the previous examples, though it was not explicitly discussed. However, `return()` can also be used to return values from any point within a function, thus ending execution of the function. The following function has an `if` statement that, when TRUE, results in early termination of the function. Notice the final result, when not true, is not even wrapped in `return()`. R returns the last object in a function by default, so an explicit call to `return()` is not strictly necessary:

```
f <- function(x) {
  if (x < 4) return("I'm done!")

  paste(x, "- Fin!")
}
f(10)

## [1] "10 - Fin!"

f(3)

## [1] "I'm done!"
```

Even though you can use return() earlier in a function, this is discouraged, because it can be surprising to users and anyone else reading or debugging code. The same effect as in the preceding code can be accomplished by using flow control:

```
f <- function(x) {
  if (x < 4) {
    "I'm done!"
  } else {
    paste(x, "- Fin!")
  }
}
f(10)

## [1] "10 - Fin!"

f(3)

## [1] "I'm done!"
```

In addition to not using return() midway in functions, some argue that an explicit call to return() should not be used at the end of functions either, as it is unnecessary. This remains a point of preference, as it can help draw attention to exactly what is returned at the end of a function.

Although not exclusively used in functions, the invisible() function is often used with the object returned by a function. Earlier we made a function that calculates the coefficient of variation and returns the function call. We can modify the function to print the coefficient of variation and invisibly return the rest, as shown in the following code. The use of invisible() means that even though the function returns the same object it did before, that object is not shown. The invisible() function is perhaps most often used in functions designed to create attractive output, such as calls to summary() or plotting functions. The function's primary purpose is to show a summary or graph, but in case anyone wants or needs to edit the object, the actual object is invisibly returned and thus can be captured and saved for later use:

```
cv <- function(x, na.rm = FALSE) {
  fcall <- match.call()

  est <- sd(x, na.rm = na.rm) / mean(x, na.rm = na.rm)

  print(est)
  return(invisible(list(CV = est, Call = fcall)))
}

cv(1:8, na.rm = TRUE)

## [1] 0.54

res <- cv(1:8, na.rm = TRUE)

## [1] 0.54

res$Call

## cv(x = 1:8, na.rm = TRUE)
```

The on.exit() function can be used to guarantee a certain set of commands is executed when the function exits or completes. Expressions in on.exit() do execute, even if the function has an error or does not properly complete as expected. An example is shown in the next use case of the little function in the following code. An error causes the function to terminate, and once that happens, the expression in on.exit() is executed. Using on.exit() is particularly valuable when a function modifies any values outside itself. For example, sometimes a plotting function modifies the default plot parameters and returns them to whatever their original state was on completion. Using on.exit() ensures that even if something goes wrong and the function fails or has an error, the user still has all of the original settings:

```
f <- function(x) {
  on.exit(print("Game over"))
  x + 5
}
```

```
f(3)
```

```
## [1] "Game over"
```

```
## [1] 8
```

```
f("a")
```

```
## Error in x + 5: non-numeric argument to binary operator
```

```
## [1] "Game over"
```

The last set of function-specific functions we cover is related to giving the software or user a signal. In order of severity, they are stop(), warning(), and message(). The next example is slightly more realistic and calculates the mean of a variable, either on its original scale or on a transformed scale, and then back-transforms the mean. The log scale can be relatively more resistant to outliers. If a log transformation is used, the example code checks whether the variable has any negative values, which are undefined and result in a full error, by using the stop() function. The argument passed to stop() is the error message to display. A similar process is followed for warning(), again with the message to be displayed in the warning. Warnings are different from errors. An error, via stop(), causes the function to stop being evaluated and terminate. A warning is issued at the end, but the function is allowed to continue its evaluation. Finally, message() can be used to send a message or signal to the user, but without indicating a real or likely problem, as a warning does. Although not demonstrated, a related convenience function is stopifnot(), which allows a logical expression to be passed and issues an error if the expression does not evaluate to a true value. Although this option has the benefit of saving some code, a disadvantage is you cannot write a custom error message to indicate exactly what went wrong.

One of the reasons it is helpful to use warnings or messages, rather than just calling print() or cat() to have the function print a message, is that warnings and messages can be (optionally) suppressed by using the aptly named functions suppressWarnings() and suppressMessages(). All of these functions are demonstrated in the code immediately following:

```r
f <- function(x, trans = c("identity", "log")) {
  trans <- match.arg(trans)

  if (trans == "log") {
    if (any(x < 0)) stop("Log is not defined for negative values")
    if (any(x < 1e-16)) warning("Some x values close or equal to zero,
    results may be unstable")

    x <- log(x)
    message("x successfully log transformed")

    exp(mean(x))
  } else {
    mean(x)
  }
}
f(c(1, 2, 100))
```

```
## [1] 34
```

```r
f(c(1, 2, 100), trans = "log")
```

```
## x successfully log transformed
```

```
## [1] 5.8
```

```r
suppressMessages(f(c(1, 2, 100), trans = "log"))
```

```
## [1] 5.8
```

```r
f(c(0, 1, 2, 100), trans = "log")
```

```
## Warning in f(c(0, 1, 2, 100), trans = "log"): Some x values close or
    equal to zero, results may be
```

```
## x successfully log transformed
```

```
## [1] 0
```

```r
suppressWarnings(f(c(0, 1, 2, 100), trans = "log"))
```

```
## x successfully log transformed
```

```
## [1] 0
```

```
f(c(-1, 1, 2, 100), trans = "log")
```

```
## Error in f(c(-1, 1, 2, 100), trans = "log"): Log is not defined for
   negative values
```

4.4 Debugging

The functions demonstrated so far have all worked or have had purposeful errors that are obvious to spot. Sometimes the process of finding the error, or debugging the code and functions written, takes longer. Fortunately, there are some tools to help the process and some practices to narrow the issues.

Although debugging can apply to any code, not just functions, functions can be particularly tricky to debug without additional tools, because normally all of their code executes without interruption or any chance to see what is happening along the way. In this section, we write a function that uses a formula to calculate the means of a variable by levels of another variable, using tapply(), which you previously examined in Chapter 3; the function then plots the raw data with larger dots for the means, using the following code, as shown in Figure 4-1.

```
meanPlot <- function(formula, d) {
  v <- all.vars(formula)
  m <- tapply(d[, v[1]], d[, v[2]],
              FUN = mean, na.rm = TRUE)

  plot(formula, data = d, type = "p")
  points(x = unique(d[, v[2]]), y = m,
         col = "blue", pch = 16, cex = 2)
}

meanPlot(mpg ~ cyl, d = mtcars)
```

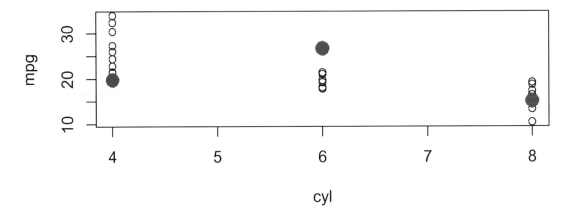

Figure 4-1. *A not yet perfected meanPlot()*

Something does not look correct about the means. The mean for the points when cyl = 8 looks okay; however, the other two seem to fall at extremes. The debug() function allows us to debug a function and step through the lines as they are executed. To use it, we first call debug() on the function we want to debug:

```
debug(meanPlot)
```

Then when we use the original function, it triggers R to enter debugging. At the first step, R tells us the function call we are debugging in, and where, by entering the debugging mode. This will change the console from > to Browse[2]>, and the commands for debugging include the ENTER (or RETURN) key to step through one bit at a time or Q to exit debugging (this can be helpful when debugging longer functions). Additionally, because debugging is moving through the operations of our function one "step" at a time, we can run code at the console to see what our variables are doing at different stages. Due to the interactive nature of debug, this is best done in real time on your own computer—although we try to explain here what you will see. We do include an example of the output you may see in Figure 4-2.

Upon running the following line of code, meanPlot(mpg ~ cyl, d = mtcars), the debug mode starts as shown in Figure 4-2 at stage 1 as indicated by debug at #1. Additionally, in RStudio, the environment window focuses on meanPlot(). Go ahead and press ENTER or RETURN at the Browse[2]> until you reach stage 7 or so where debug at #7: points(x = unique(d[, v[2]]), y = m, col = "blue", pch = 16, cex = 2) occurs—this is where the mean points are added to our plot. At that stage, check the variables going into the point plot. This is done by running Browse[2]> *unique(d[, v[2]])*. This shows the order of the x-axis variables is wrong! After seeing this mistake in our function's code, enter a capital Q into the debugging browser to quit the debugging mode:

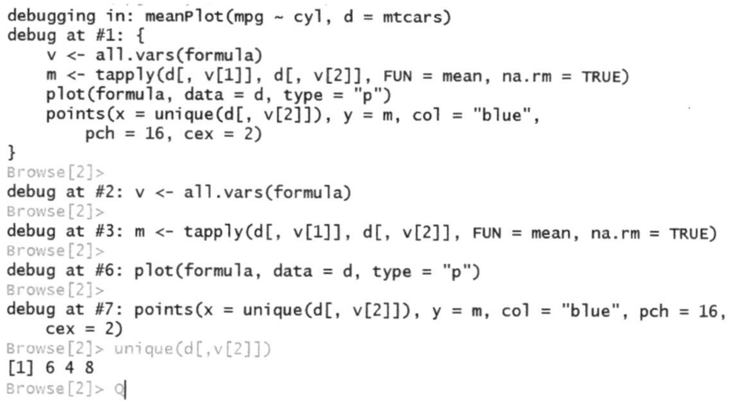

```
debugging in: meanPlot(mpg ~ cyl, d = mtcars)
debug at #1: {
    v <- all.vars(formula)
    m <- tapply(d[, v[1]], d[, v[2]], FUN = mean, na.rm = TRUE)
    plot(formula, data = d, type = "p")
    points(x = unique(d[, v[2]]), y = m, col = "blue",
        pch = 16, cex = 2)
}
Browse[2]>
debug at #2: v <- all.vars(formula)
Browse[2]>
debug at #3: m <- tapply(d[, v[1]], d[, v[2]], FUN = mean, na.rm = TRUE)
Browse[2]>
debug at #6: plot(formula, data = d, type = "p")
Browse[2]>
debug at #7: points(x = unique(d[, v[2]]), y = m, col = "blue", pch = 16,
    cex = 2)
Browse[2]> unique(d[,v[2]])
[1] 6 4 8
Browse[2]> Q
```

Figure 4-2. *A screenshot of debug()*

```
meanPlot(mpg ~ cyl, d = mtcars
```

We know now where the problem occurs and have exited debug mode by using Q + ENTER. We fix our function by first sorting the data, so tapply() and unique() give the results in the same order. The code is shown here, and the result is plotted in Figure 4-3.

```
meanPlot <- function(formula, d) {
  v <- all.vars(formula)
  d <- d[order(d[, v[2]]), ] #This line is the fix by sorting first!
  m <- tapply(d[, v[1]], d[, v[2]],
            FUN = mean, na.rm = TRUE)

  plot(formula, data = d, type = "p")
  points(x = unique(d[, v[2]]), y = m,
        col = "blue", pch = 16, cex = 2)
}

meanPlot(mpg ~ cyl, d = mtcars)
```

If you are debugging your function and do not want to step through each line of code, the browser() function can be inserted into the function code. Then, when the function reaches that point, a browser is invoked, and you can examine the current state of variables in the function's local environment. In the following example code, we add a call to browser() in the function. Once running meanPlot(mpg cyl, d = mtcars), in the subsequent browser, we use ENTER once and then examine the current objects available by using ls(), followed by viewing the contents of the object, v, and finally type a capital Q to quit debugging. Because the browser is interactive, we show the approximate output of your console in Figure 4-4.

```
meanPlot <- function(formula, d) {
  v <- all.vars(formula)
  d <- d[order(d[, v[2]]), ] ## sorting first
  m <- tapply(d[, v[1]], d[, v[2]],
              FUN = mean, na.rm = TRUE)

  browser()

  plot(formula, data = d, type = "p")
  points(x = unique(d[, v[2]]), y = m,
         col = "blue", pch = 16, cex = 2)
}

meanPlot(mpg ~ cyl, d = mtcars)
```

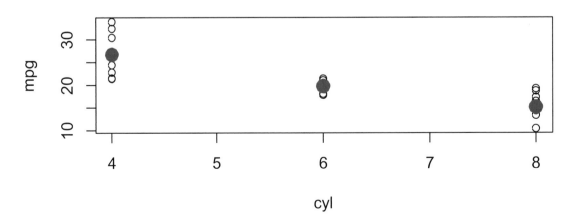

Figure 4-3. *A now perfected meanPlot()*

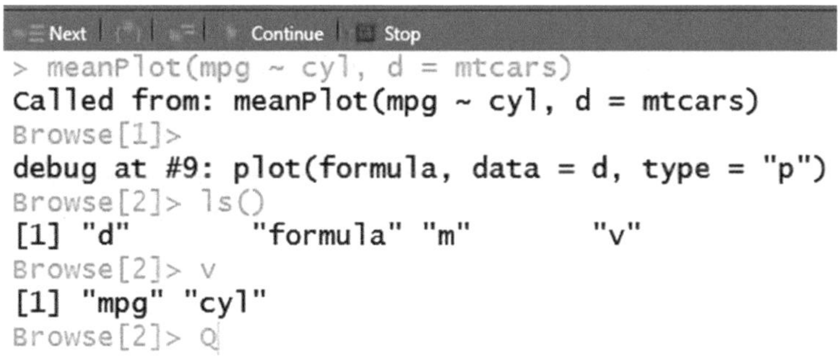

```
  ≡ Next              Continue     ⏹ Stop
> meanPlot(mpg ~ cyl, d = mtcars)
Called from: meanPlot(mpg ~ cyl, d = mtcars)
Browse[1]>
debug at #9: plot(formula, data = d, type = "p")
Browse[2]> ls()
[1] "d"         "formula" "m"         "v"
Browse[2]> v
[1] "mpg" "cyl"
Browse[2]> Q
```

Figure 4-4. *A screenshot of browser()*

A useful function for debugging when working interactively is `traceback()`. For example, if you call a function such as `lm()` and then get an error message, it can sometimes be difficult to know exactly where or why that mistake occurred. The error is often not even directly from the function you called, as the functions you frequently use may in turn call many other functions internally. The following example shows how this is done by using `traceback()` immediately after the code that resulted in error. The output shows the call stack tracing the path from the initial call to the final code that generated the error. This can be useful information if you need to look at the code to determine why the error occurred:

```
lm(mpg ~ jack, data = mtcars)

## Error in eval(predvars, data, env): object 'jack' not found

traceback()
## 7: eval(predvars, data, env)
## 6: eval(predvars, data, env)
## 5: model.frame.default(formula = mpg ~ jack, data = mtcars, drop.unused.
       levels = TRUE)
## 4: stats::model.frame(formula = mpg ~ jack, data = mtcars, drop.unused.
       levels = TRUE)
## 3: eval(mf, parent.frame())
## 2: eval(mf, parent.frame())
## 1: lm(mpg ~ jack, data = mtcars)
```

In this case, the error happens because `jack` is not a column name of `mtcars`.

Finally, sometimes code bugs are not in your code, but in other code you are using, such as from another R package. Although it is rare to find bugs in recommended R packages, it is more frequent in the thousands of other R packages. Also, sometimes problems are not a bug per se, but a difference in how you want to use a function vs. how the original writer envisioned its use. Although the source code for all R packages on CRAN is publicly available for download and editing, it can be a hassle to download an entire package's source code, edit, and reinstall, just to see if doing something slightly different in one function fixes the problem.

Consider the following challenge. Suppose you are using data that sometimes includes infinity for some reason, yet you want to include only finite cases. The following code shows an example using the `wtd.quantile()` function from the `Hmisc` package:

```
wtd.quantile(c(1, 2, 3, Inf, NA),
        weights = c(.6, .9, .4, .2, .6))
```

```
##    0%   25%  50%  75%  100%
## NaN  Inf  Inf  Inf   Inf
```

Looking at the code for `wtd.quantile()`, we see another function does the main calculations, `wtd.table()`, as shown here:

```
wtd.quantile
```

```
## function (x, weights = NULL, probs = c(0, 0.25, 0.5, 0.75, 1),
##     type = c("quantile", "(i-1)/(n-1)", "i/(n+1)", "i/n"), normwt = FALSE,
##     na.rm = TRUE)
## {
##     if (!length(weights))
##         return(quantile(x, probs = probs, na.rm = na.rm))
##     type <- match.arg(type)
##     if (any(probs < 0 | probs > 1))
##         stop("Probabilities must be between 0 and 1 inclusive")
##     nams <- paste(format(round(probs * 100, if (length(probs) >
##         1) 2 - log10(diff(range(probs))) else 2)), "%", sep = "")
##     i <- is.na(weights) | weights == 0
##     if (any(i)) {
##         x <- x[!i]
```

```
##          weights <- weights[!i]
##       }
##       if (type == "quantile") {
##          w <- wtd.table(x, weights, na.rm = na.rm, normwt = normwt,
##              type = "list")
##          x <- w$x
##          wts <- w$sum.of.weights
##          n <- sum(wts)
##          order <- 1 + (n - 1) * probs
##          low <- pmax(floor(order), 1)
##          high <- pmin(low + 1, n)
##          order <- order%%1
##          allq <- approx(cumsum(wts), x, xout = c(low, high), method =
##          "constant",
##              f = 1, rule = 2)$y
##          k <- length(probs)
##          quantiles <- (1 - order) * allq[1:k] + order * allq[-(1:k)]
##          names(quantiles) <- nams
##          return(quantiles)
##       }
##       w <- wtd.Ecdf(x, weights, na.rm = na.rm, type = type, normwt = normwt)
##       structure(approx(w$ecdf, w$x, xout = probs, rule = 2)$y,
##          names = nams)
## }
## <bytecode: 0x000002ac904e4750>
## <environment: namespace:Hmisc>
```

Using the same approach, we examine the wtd.table() function. When doing so, we see that although it can automatically remove missing values, it has no check or way to remove nonfinite values:

```
wtd.table
```

```
## function (x, weights = NULL, type = c("list", "table"), normwt = FALSE,
##     na.rm = TRUE)
## {
##        type <- match.arg(type)
```

```
##      if (!length(weights))
##          weights <- rep(1, length(x))
##      isdate <- testDateTime(x)
##      ax <- attributes(x)
##      ax$names <- NULL
##      if (is.character(x))
##          x <- as.factor(x)
##      lev <- levels(x)
##      x <- unclass(x)
##      if (na.rm) {
##          s <- !is.na(x + weights)
##          x <- x[s, drop = FALSE]
##          weights <- weights[s]
##      }
##      n <- length(x)
##      if (normwt)
##          weights <- weights * length(x)/sum(weights)
##      i <- order(x)
##      x <- x[i]
##      weights <- weights[i]
##      if (anyDuplicated(x)) {
##          weights <- tapply(weights, x, sum)
##          if (length(lev)) {
##              levused <- lev[sort(unique(x))]
##              if ((length(weights) > length(levused)) && any(is.na(weights)))
##                  weights <- weights[!is.na(weights)]
##              if (length(weights) != length(levused))
##                  stop("program logic error")
##              names(weights) <- levused
##          }
##          if (!length(names(weights)))
##              stop("program logic error")
##          if (type == "table")
##              return(weights)
##          x <- all.is.numeric(names(weights), "vector")
```

```
##            if (isdate)
##                attributes(x) <- c(attributes(x), ax)
##            names(weights) <- NULL
##            return(list(x = x, sum.of.weights = weights))
##        }
##        xx <- x
##        if (isdate)
##            attributes(xx) <- c(attributes(xx), ax)
##        if (type == "list")
##            list(x = if (length(lev)) lev[x] else xx, sum.of.weights = weights)
##        else {
##            names(weights) <- if (length(lev))
##                lev[x]
##            else xx
##            weights
##        }
## }
## <bytecode: 0x000002ac904f2cc8>
## <environment: namespace:Hmisc>
```

It may seem easier just to change your data, but sometimes data is generated automatically and passed on, so it may be simpler to change a function than to modify the data. We can readily copy and paste the code for wtd.table() into our editor and revise, as shown here:

```
revised.wtd.table <- function (x, weights = NULL, type = c("list",
"table"), normwt = FALSE, na.rm = TRUE)
{
    type <- match.arg(type)
    if (!length(weights))
        weights <- rep(1, length(x))
    isdate <- testDateTime(x)
    ax <- attributes(x)
    ax$names <- NULL
    if (is.character(x))
        x <- as.factor(x)
    lev <- levels(x)
```

```
x <- unclass(x)
if (na.rm) {
    s <- !is.na(x + weights) & is.finite(x + weights) ##here is the fix
    x <- x[s, drop = FALSE]
    weights <- weights[s]
}
n <- length(x)
if (normwt)
    weights <- weights * length(x)/sum(weights)
i <- order(x)
x <- x[i]
weights <- weights[i]
if (anyDuplicated(x)) {
    weights <- tapply(weights, x, sum)
    if (length(lev)) {
        levused <- lev[sort(unique(x))]
        if ((length(weights) > length(levused)) && any(is.na(weights)))
            weights <- weights[!is.na(weights)]
        if (length(weights) != length(levused))
            stop("program logic error")
        names(weights) <- levused
    }
    if (!length(names(weights)))
        stop("program logic error")
    if (type == "table")
        return(weights)
    x <- all.is.numeric(names(weights), "vector")
    if (isdate)
        attributes(x) <- c(attributes(x), ax)
    names(weights) <- NULL
    return(list(x = x, sum.of.weights = weights))
}
xx <- x
if (isdate)
    attributes(xx) <- c(attributes(xx), ax)
```

```
    if (type == "list")
        list(x = if (length(lev)) lev[x] else xx, sum.of.weights = weights)
    else {
        names(weights) <- if (length(lev))
            lev[x]
        else xx
        weights
    }
}
```

Unlike the original wtd.table() function, the revised function works exactly as we want:

```
wtd.table(c(1, 2, 3, Inf, NA),
        weights = c(.6, .9, .4, .2, .6))
```

```
## $x
## [1]    1    2    3    Inf
##
## $sum.of.weights
## [1] 0.6 0.9 0.4 0.2
```

```
revised.wtd.table(c(1, 2, 3, Inf, NA),
        weights = c(.6, .9, .4, .2, .6))
```

```
## $x
## [1] 1 2 3
##
## $sum.of.weights
## [1] 0.6 0.9 0.4
```

However, the challenge is the wtd.quantile() function; and other Hmisc functions using wtd.table() still do not work as we hope, because they do not use our revised function. Assigning our function to the name wtd.table() in the global environment is not sufficient, because scoping rules mean that Hmisc functions access the wtd.table() function first from the Hmisc environment, not from our global environment. It is like having a file of the same name on your computer, but in two different folders. For all the Hmisc functions that utilize wtd.table() to use our revised function, we need not only to name it correctly but also to put it in the correct place. We can assign an object to a specific namespace by using the assignInNamespace() function:

```
assignInNamespace(x = "wtd.table", #uses expected name in all other Hmisc
functions
                  value = revised.wtd.table, #force our modified function
                  ns = "Hmisc") #in the NameSpace of Hmisc
wtd.quantile(c(1, 2, 3, Inf, NA),
        weights = c(.6, .9, .4, .2, .6))

##   0%  25%  50%  75% 100%
##  2.0  2.2  2.4  2.7  2.9

wtd.Ecdf(c(1, 2, 3, Inf, NA),
        weights = c(.6, .9, .4, .2, .6))

## $x
## [1] 1 1 2 3
##
## $ecdf
## [1] 0.00 0.32 0.79 1.00
```

Although the assignInNamespace() function has no output, we can see that afterward the wtd.quantile() function and others that depend on wtd.table(), such as wtd.Ecdf(), are now working as we hoped. A caveat about assignInNamespace() is you cannot assign any object to an object that does not already exist in that namespace (i.e., you can overwrite only existing objects). You also are not allowed to use assignInNamespace() in any packages you may want to submit to CRAN (more on that in Chapter 6). Even if it is sometimes the convenient approach, it would become confusing if people did this as a general rule. Were this to happen, the definition of functions in package A would depend on whether you had loaded package B, because package B could overwrite (not just mask) the function definition in package A. Note that any changes you make in this way are temporary and vanish when you restart R. However, this can be a helpful technique for debugging an existing package, as it is a relatively easy way to make sure the issue encountered using wtd.quantile() was indeed "fixed" by the suggested change to wtd.table(). At this point, if it were truly a bug or even if it was just a desirable feature, you could email the package maintainer to suggest the change, confident your suggested code works. To see the current maintainer, run maintainer("Hmisc"), substituting whatever the package of interest is called for Hmisc.

4.5 Summary

We covered many functions and much about functions in this chapter! In case you need to refresh your memory about any specific functions, Table 4-1 lists the key functions introduced in this chapter and provides a brief description of each. Of course, you can look up more in the official help files.

Although the formal names of various parts of a function are not necessarily critical, learning how to use and write functions efficiently may be one of the best investments in learning R you ever make. Writing functions provides a way out of writing repetitive code. There are no strict rules, but if you find yourself doing the same task often, there is a good chance it is worth writing a function to do that. It might be a big task involving many pieces, or it might be a small task. For example, in psychology, it is common to report "high" and "low" values of a continuous variable, which are often defined as mean +/– 1 standard deviation. This is easy to do in R by using the mean() and sd() functions. If you do it a lot, it may be worth writing a short, one-line function so that rather than type mean(x) + sd(x), you just type msd(x) or whatever you call your function. If you do not feel comfortable playing with functions and writing your own, it would be a good idea to get some more practice before moving on to Chapters 5 and 6, where we assume you are comfortable with functions.

Table 4-1. *Listing of key functions described in this chapter and summary of what they do*

Function	What It Does
formals()	Allows you to see the formal arguments of a function you build.
args()	Shows default values and names for a created function's arguments.
body()	Shows the body of a function.
environment()	Shows the environment your function lives in (often the global environment, so far).
search()	Shows the search order for functions. Remember, people may have already used your function name!
parent.env()	Takes an environment and tracks it up one level.
match.arg()	Allows for fuzzy matching of function arguments.

(continued)

Table 4-1. (*continued*)

Function	What It Does
`missing()`	Tests whether a value was passed to the function. Note that this will be false after `match.arg()`.
`match.call()`	Captures the entire function call; used often in regression.
`return()`	Not required, possibly contentious, and for sure, if used be the last part of a new function.
`invisible()`	Suppresses output.
`on.exit()`	Regardless of a successful or failed function attempt, this executes its argument.
`stop()` , `stopifnot()`, `warning()`, `message()`	These are errors of various levels of severity, from full-on stop and an error to a milder warning to a mostly polite message.
`suppressWarnings()`, `suppressMessages()`	These two do precisely what they say. Once you're familiar with a function, warnings or messages may be tedious or safely ignored.
`debug()`	Use this to start debugging a function that is not working properly.
`browser()`	This goes into a function body, right before the part you want to debug.
`traceback()`	Provides a list of expressions leading to the source of an error.
`assignInNamespace()`	Allows for the temporary replacement of a function with a locally crafted function.
`maintainer()`	Called on a package, shows a contact name and email to go to for troubleshooting.

CHAPTER 5

Writing Classes and Methods

It is often helpful to have a function behave differently depending on the type of object passed. For example, when summarizing a variable, it makes sense to create a different summary for numeric or string data. It is possible to have a different function for every type of object, but then users would have to remember many function names, and to remain unique, function names may be longer. Object-oriented programming (OOP) is based on objects and is implemented in R (as in most programming languages) by using two concepts: classes and methods. A class defines a template, or blueprint, describing the variables and features of an object as well as determining what methods work for it. For example, a house may be defined as having a floor, four walls, a roof, and a door. Specific data represents these properties, such as the dimensions and color of each wall. The methods are behaviors or actions that can be performed on a particular object type. For instance, a house can be painted, which changes its color, but a house cannot be eaten. R has three object-oriented systems: S3, S4, and R5. This chapter covers the S3 and S4 systems, which are the most common.

In this chapter, we use the ggplot2 package [25]. The following code loads the ggplot2 package and sets some options for text display:

```
options(width = 70, digits = 2)
library(ggplot2)
```

© Matt Wiley and Joshua F. Wiley 2020
M. Wiley and J. F. Wiley, *Advanced R 4 Data Programming and the Cloud*,
https://doi.org/10.1007/978-1-4842-5973-3_5

5.1 S3 System

The S3 system is the most common object-oriented system. It is also the easiest to start using and the simplest of the systems. The S3 system is easy to use in part because it is quite informal and mostly focused on the functions or methods. These advantages are also limitations, as the S3 system provides no formal framework for ensuring that objects meet the requirements for a class. For a more in-depth guide to R programming using the S3 system, see *S Programming* by W. N. Venables and B. D. Ripley (Springer, 2004).

S3 Classes

In R, some types, or classes, of objects are available by default, and almost all can be thought of as vectors or generic vectors. For example, matrices and arrays are essentially vectors with attributes indicating the dimensions. Lists are generic vectors, in which each element of the vector may contain another vector. Data frames are lists in which each item is a vector, but all with equal length, and thus have a tabular format. Vectors can hold specific types of data, such as logical, integer, numeric (real numbers), or character strings. These fundamental objects and classes are provided by the base package. S3 classes are created by building on the objects and classes provided by the base package.

Note   S3 classes are created by using regular R objects (e.g., vectors or lists) and classes (e.g., logical or numeric). S3 classes are defined by setting the class name via class(object) <- "class name". Elements of S3 objects are typically accessed by using $, [, or [[.

The practical creation of S3 classes is simple. First, create an R object that meets the requirements or characteristics of the class, and then define the S3 classes by labeling the object with the class name. Because S3 classes differ from other R objects only in having special names or attributes, they are accessed and manipulated the same way as other basic R objects, by using $, [, or [[. To check the class of an object, we can use the class() function. In the following example, the mtcars object is queried to determine that it has a data.frame class:

```r
#### S3 System - Classes ####
class(mtcars)

## [1] "data.frame"
```

To see how the class of an object is set, we can look at many of the functions from base R, such as table(). Here, we print the source code for the table() function, leaving some of the middle off to save space:

```r
table

## function (..., exclude = if (useNA == "no") c(NA, NaN), useNA = c("no",
##     "ifany", "always"), dnn = list.names(...), deparse.level = 1)
## {
##     list.names <- function(...) {
##         l <- as.list(substitute(list(...)))[-1L]
##         nm <- names(l)
##         fixup <- if (is.null(nm))
##             seq_along(l)
##         else nm == ""
##         dep <- vapply(l[fixup], function(x) switch(deparse.level +
##             1, "", if (is.symbol(x)) as.character(x) else "",
##             deparse(x, nlines = 1)[1L]), "")
##         if (is.null(nm))
##             dep
##         else {
##             nm[fixup] <- dep
##             nm
##         }
##     }
##     miss.use <- missing(useNA)
##     miss.exc <- missing(exclude)
##     useNA <- if (miss.use && !miss.exc && !match(NA, exclude,
##         nomatch = 0L))
##         "ifany"
##     else match.arg(useNA)
##     doNA <- useNA != "no"
```

```
##        if (!miss.use && !miss.exc && doNA && match(NA, exclude,
##            nomatch = 0L))
##            warning("'exclude' containing NA and 'useNA' != \"no\"' are a
            bit contradicting")
##      args <- list(...)
##      if (!length(args))
##          stop("nothing to tabulate")
##          if (length(args) == 1L && is.list(args[[1L]])) {
##          args <- args[[1L]]
##          if (length(dnn) != length(args))
##              dnn <- if (!is.null(argn <- names(args)))
##                  argn
##              else paste(dnn[1L], seq_along(args), sep = ".")
##      }
##      bin <- 0L
##      lens <- NULL
##      dims <- integer()
##      pd <- 1L
##      dn <- NULL
##      for (a in args) {
##          if (is.null(lens))
##              lens <- length(a)
##          else if (length(a) != lens)
##              stop("all arguments must have the same length")
##          fact.a <- is.factor(a)
##          if (doNA)
##              aNA <- anyNA(a)
##          if (!fact.a) {
##              a0 <- a
##              a <- factor(a, exclude = exclude)
##          }
##          add.na <- doNA
##          if (add.na) {
##              ifany <- (useNA == "ifany")
##              anNAc <- anyNA(a)
```

```
##            add.na <- if (!ifany || anNAc) {
##                ll <- levels(a)
##                if (add.ll <- !anyNA(ll)) {
##                   ll <- c(ll, NA)
##                   TRUE
##                }
##                else if (!ifany && !anNAc)
##                   FALSE
##                else TRUE
##            }
##            else FALSE
##         }
##         if (add.na)
##             a <- factor(a, levels = ll, exclude = NULL)
##         else ll <- levels(a)
##         a <- as.integer(a)
##         if (fact.a && !miss.exc) {
##             ll <- ll[keep <- which(match(ll, exclude, nomatch = 0L) ==
##                 0L)]
##             a <- match(a, keep)
##         }
##         else if (!fact.a && add.na) {
##             if (ifany && !aNA && add.ll) {
##                 ll <- ll[!is.na(ll)]
##                 is.na(a) <- match(a0, c(exclude, NA), nomatch = 0L) >
##                     0L
##             }
##             else {
##                 is.na(a) <- match(a0, exclude, nomatch = 0L) >
##                     0L
##             }
##         }
##         nl <- length(ll)
##         dims <- c(dims, nl)
##         if (prod(dims) > .Machine$integer.max)
```

```
##                  stop("attempt to make a table with >= 2^31 elements")
##          dn <- c(dn, list(ll))
##          bin <- bin + pd * (a - 1L)
##          pd <- pd * nl
##      }
##      names(dn) <- dnn
##      bin <- bin[!is.na(bin)]
##      if (length(bin))
##          bin <- bin + 1L
##      y <- array(tabulate(bin, pd), dims, dimnames = dn)
##      class(y) <- "table"
##      y
## }
## <bytecode: 0x000002ac81ef2848>
## <environment: namespace:base>
```

The object returned at the end, y, is an array, but has a table class. The object class is set using the idiom class(object) <- "class name". It is also possible to assign more than one class to an object. This is done in much the same fashion as assigning a single class and in order of preference. The following example stores the results from calling table() in the object x and then sets two classes, first newclass and then the original table class:

```
x <- table(mtcars$cyl)
class(x) <- c("newclass", "table")
class(x)
```

```
## [1] "newclass" "table"
```

The value of assigning multiple classes is a sort of backup, primarily for methods. For example, if you create a new class that is a variant of a table or data frame, you may write a dedicated method for printing, yet depend on methods of the original class for other functions, such as plotting or summaries.

The types of classes you can write are virtually endless. A simple way to start is by creating special or augmented cases of existing classes. In the following example, we make a special case of a data.frame, with x and y coordinates and text labels, called textplot. The simplest way to "create" the class is to create a data.frame and then change its class. If we use only our new class label, calling the print() function results in the default method, but if we label it with both our new textplot label and as a data.frame secondly, then calling print() falls back to the method for data.frames:

```
d <- data.frame(
  x = c(1, 3, 5),
  y = c(1, 2, 4),
  labels = c("First", "Second", "Third")
)
class(d) <- "textplot"
```

print(d)

```
## $x
## [1] 1 3 5
##
## $y
## [1] 1 2 4
##
## $labels
## [1] "First" "Second" "Third"
##
## attr(,"class")
## [1] "textplot"
## attr(,"row.names")
## [1] 1 2 3
```

```
class(d) <- c("textplot", "data.frame")
```

print(d)

```
##   x y labels
## 1 1 1  First
## 2 3 2 Second
## 3 5 4  Third
```

In the S3 system, there is no formal way to create an object from a particular class. However, the most common way to create objects of a specific class is as the output from a function. Dedicated functions can be written to create an object of a specific class, or functions can be written that perform operations and output results as an object of a specific class. The latter approach is more typical in R when using the S3 system.

When you are creating a new class, some of the desired features or elements may be present in an existing class. If this is the case, it may make sense to build on, or extend, an existing class. For instance, data frames build on lists, requiring that each element of the list be a vector with the same length. Likewise, the textplot class we created previously builds on data frames, requiring three elements named x, y, and labels. In this instance, we would say that textplot *inherits* from data.frame. That is, textplot is a child of the parent class, data.frame (note that parent classes are also referred to as the *super* class, or *base* class). If a class inherits from only one other class, it is called single inheritance (i.e., it has only one parent class). If a class inherits from multiple classes, it is called multiple inheritance (i.e., it has more than one parent class). If a class inherits from another class, it may inherit features of the data stored, and it may inherit methods, the functions that operate on objects of a specific class. Inheriting methods is especially useful to effort duplication.

Note   Inheritance refers to creating a new class by building on the features and methods of an existing class. The new class is known as the child class, and the classes from which the new class is derived are the parent, or super, classes. A child class may inherit both features of the parent class and use of the parent class' methods.

In the next example, we write a function to use the formula interface to build a textplot object from a data frame. The function has two arguments: the formula, called f, and the data, called d. The first part of the code uses the stopifnot() function introduced in Chapter 4 to check that the classes of the objects passed to the function match what is expected. To test the classes, we use the inherits() function, which assesses whether a particular object is, or inherits from, the specified class. For example, our textplot class is secondly a data.frame, and so testing that uses inherits(object, "data.frame") would evaluate to TRUE. When we build S4 classes, we see a more formal definition and system for class inheritance. The next part of the function gets all the variables from the formula, in order, from the specified data frame, renames the columns, applies our textplot class, and returns the object:

```
textplot_data <- function(f, d) {
  stopifnot(inherits(d, "data.frame"))
  stopifnot(inherits(f, "formula"))

  newdata <- get_all_vars(formula = f, data = d)
  colnames(newdata) <- c("y", "x", "labels")
  class(newdata) <- c("textplot", "data.frame")

return(newdata)
}

textplot_data(f = mpg ~ hp | cyl, d = mtcars[1:10, ])
```

```
##                    y   x labels
## Mazda RX4          21 110      6
## Mazda RX4 Wag      21 110      6
## Datsun 710         23  93      4
## Hornet 4 Drive     21 110      6
## Hornet Sportabout  19 175      8
## Valiant            18 105      6
## Duster 360         14 245      8
## Merc 240D          24  62      4
## Merc 230           23  95      4
## Merc 280           19 123      6
```

These examples show how easy it is to use the S3 system to create classes. Next, we explore how to write methods for existing or new classes.

S3 Methods

Methods are functions or operations that can be performed on objects of specific classes. Even if you do not write your own classes, you may write your own methods. Writing S3 methods is like writing functions as we did in Chapter 4. The only difference is S3 methods have a special naming convention and require a generic function for users, which takes care of dispatching to the appropriate method.

Note   S3 methods are regular R functions that follow a specific naming convention: `FunctionVerb.classname()`. Users call the generic function, `FunctionVerb()`, which dispatches to the appropriate method based on the class of object passed in as an argument. If no generic function exists, a generic must be written that includes a call to `UseMethods()`, which handles the actual method dispatch.

To start, we write a simple plotting method for the `textplot` object class we developed. To make an S3 plot method, we use the function verb `plot`, followed by the class name `textplot` which in this case is `plot.textplot()`. As long as we name our function in this way, it works as an S3 method, as long as a generic `plot()` function exists, a topic we discuss shortly.

Our plot function is shown in the following code. The call to the `par()` function adjusts the default margins for a graph, to reduce excess white space on the top and right of our graph. The results are stored in the object, `op`, as this stores the original graphical parameters. Then the user's original graphical parameter state can be restored when the function exits by calling `par(op)` on exit, a function you learned about in Chapter 4. Next, we create a new plot area by calling `plot.new()` and set the dimensions of our new plot by calling `plot.window()` with the x and y limits determined by the range of the data. Next, we plot the labels by using the `text()` function, which takes the coordinates and labels. Finally, we add an axis on the bottom, `side = 1`, and left, `side = 2`. The original `textplot` data object is returned invisibly at the end:

```
#### S3 System - Methods ####

plot.textplot <- function(d) {
  ## adjust the margins through a call to the
  ## graphical parameters, and store old parameters in op
  op <- par(mar = c(4, 4, 1, 1))
  ## ensure graphing parameters restored to what
  ## they were when function completes
  on.exit(par(op))
  ## create a new plot area
  plot.new()
  ## size the plot appropriate for the data
```

```
plot.window(xlim = range(d$x, na.rm = TRUE),
            ylim = range(d$y, na.rm = TRUE))
## add the text to the plot
text(d$x, d$y, labels = d$labels)
## create some axes to show the limits of the data
axis(side = 1, range(d$x, na.rm = TRUE))
axis(side = 2, range(d$y, na.rm = TRUE))
## return the object invisibly
invisible(d)
}
```

Next, we need to make some data and then plot() it. Note that because it is a method, we do not need to call our function by its full name, plot.textplot(). We can simply call plot(), and R takes care of dispatching the data object to the correct method based on the object class, textplot. The result is shown in Figure 5-1, and the code is shown here:

```
dat <- textplot_data(f = mpg ~ hp | cyl, d = mtcars[1:10, ])

plot(dat)
```

This example shows just how easy it is to use the S3 system. Almost no special functions or effort is required. Just create a regular R object however you want, add a custom class label, write a function, and give it a special name. That is essentially all that is required. However, there are a few special functions and tools for S3 methods. To start, let us see what happens when we look at the source code for plot(). This generic function contains just three arguments.

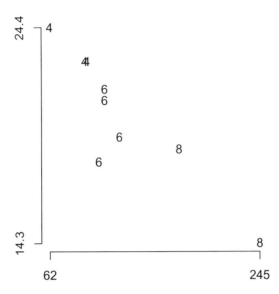

Figure 5-1. *The plot of text labels at specific coordinates, demonstrating the use of custom methods for custom classes*

The body consists only of a call to UseMethod(), which is what tells R to check the argument classes and dispatch to an appropriate method. If you are writing methods and a generic function does not exist, you also need to write the generic function. Writing a generic function in the S3 system is a straightforward task that requires only considering the default arguments to include:

plot

[1] 5

To see the methods available, we use the methods() function. The result shows many specific methods available for plot(), including our newly written plot.textplot:

methods(**plot**)

```
##  [1] plot,ANY-method         plot,color-method
##  [3] plot.aareg*             plot.acf*
##  [5] plot.agnes*             plot.areg*
##  [7] plot.areg.boot*         plot.aregImpute*
##  [9] plot.biVar*             plot.clusGap*
## [11] plot.cox.zph*           plot.curveRep*
## [13] plot.data.frame*        plot.decomposed.ts*
```

```
## [15] plot.default                    plot.dendrogram*
## [17] plot.density*                    plot.describe*
## [19] plot.diana*                      plot.drawPlot*
## [21] plot.ecdf                        plot.factor*
## [23] plot.formula*                    plot.function
## [25] plot.gbayes*                     plot.ggplot*
## [27] plot.gtable*                     plot.hcl_palettes*
## [29] plot.hclust*                     plot.histogram*
## [31] plot.HoltWinters*                plot.isoreg*
## [33] plot.lm*                         plot.medpolish*
## [35] plot.mlm*                        plot.mona*
## [37] plot.partition*                  plot.ppr*
## [39] plot.prcomp*                     plot.princomp*
## [41] plot.profile.nls*                plot.Quantile2*
## [43] plot.R6*                         plot.raster*
## [45] plot.rm.boot*                    plot.rpart*
## [47] plot.shingle*                    plot.silhouette*
## [49] plot.spec*                       plot.spline*
## [51] plot.stepfun                     plot.stl*
## [53] plot.summary.formula.response*   plot.summary.formula.reverse*
## [55] plot.summaryM*                   plot.summaryP*
## [57] plot.summaryS*                   plot.Surv*
## [59] plot.survfit*                    plot.table*
## [61] plot.textplot                    plot.trans*
## [63] plot.transcan*                   plot.trellis*
## [65] plot.ts                          plot.tskernel*
## [67] plot.TukeyHSD*                   plot.varclus*
## [69] plot.xyVector*
## see '?methods' for accessing help and source code
```

Some of the functions are followed by an asterisk, indicating that these methods are not public or cannot be directly accessed, such as plot.table(). For example, if we type its name, we get an error that it cannot be found:

```
plot.table
```

```
## Error in eval(expr, envir, enclos): object 'plot.table' not found
```

> **Note**   The :: operator is used to refer to publicly exported functions from a specific package, primarily when multiple packages export functions with the same name: `package::FunctionVerb()`. The ::: operator is used to access functions from a package's namespace that are not exported. Use nonpublic functions with caution, as they are subject to change without notice.

Functions that are not public or have not been exported from a package namespace can still be accessed as methods. These functions also can be accessed directly by specifying the package they are from and using the ::: operator, revealing the function source code:

```r
graphics:::plot.table
## function (x, type = "h", ylim = c(0, max(x)), lwd = 2, xlab = NULL,
##     ylab = NULL, frame.plot = is.num, ...)
## {
##     xnam <- deparse1(substitute(x))
##     rnk <- length(dim(x))
##     if (rnk == 0L)
##         stop("invalid table 'x'")
##     if (rnk == 1L) {
##         dn <- dimnames(x)
##         nx <- dn[[1L]]
##         if (is.null(xlab))
##             xlab <- names(dn)
##         if (is.null(xlab))
##             xlab <- ""
##         if (is.null(ylab))
##             ylab <- xnam
##         is.num <- suppressWarnings(!any(is.na(xx <- as.numeric(nx))))
##         x0 <- if (is.num)
##             xx
##         else seq_along(x)
##         plot(x0, unclass(x), type = type, ylim = ylim, xlab = xlab,
```

```
##               ylab = ylab, frame.plot = frame.plot, lwd = lwd,
##               ..., xaxt = "n")
##          localaxis <- function(..., col, bg, pch, cex, lty, log) axis(...)
##          if (!isFALSE(list(...)$axes))
##              localaxis(1, at = x0, labels = nx, ...)
##      }
##      else {
##          if (length(dots <- list(...)) && !is.null(dots$main))
##              mosaicplot(x, xlab = xlab, ylab = ylab, ...)
##          else mosaicplot(x, xlab = xlab, ylab = ylab, main = xnam,
##              ...)
##      }
## }
## <bytecode: 0x000002ac901d97f8>
## <environment: namespace:graphics>
```

In addition to writing methods for new classes, it is sometimes helpful to write methods for existing classes. For example, the popular ggplot2 package has no default method for working with a linear model or regression objects. To start, we set up a simple regression model by using the mtcars data. From there, we are predicting mpg from hp, vs, their interaction (all of which are created from hp * vs, which expands to the two main effects and their interaction or product term), and cyl dummy coded, through the call to factor(). The results are shown here by calling summary() on the object:

```
m <- lm(mpg ~ hp * vs + factor(cyl), data = mtcars)
summary(m)

##
## Call:
## lm(formula = mpg ~ hp * vs + factor(cyl), data = mtcars)
##
## Residuals:
##    Min    1Q Median    3Q    Max
## -4.76  -1.44  -0.17  1.59   6.94
##
## Coefficients:
##              Estimate Std. Error t value Pr(>|t|)
```

```
## (Intercept)   26.9291    2.7176    9.91  2.6e-10 ***
## hp            -0.0152    0.0155   -0.98  0.3372
## vs             8.5335    4.9530    1.72  0.0968 .
## factor(cyl)6  -4.2112    1.9389   -2.17  0.0392 *
## factor(cyl)8  -8.6510    2.6974   -3.21  0.0035 **
## hp:vs         -0.0910    0.0436   -2.09  0.0469 *
## ---
## Signif. codes:  0 '***' 0.001 '**' 0.01 '*' 0.05 '.' 0.1 ' ' 1
##
## Residual standard error: 3 on 26 degrees of freedom
## Multiple R-squared:  0.791,  Adjusted R-squared:  0.751
## F-statistic: 19.7 on 5 and 26 DF,  p-value: 4.18e-08
```

We can check the class of the object and that there is no current `ggplot()` method by using the following code:

```
class(m)
```

```
## [1] "lm"
```

```
methods(ggplot)
```

```
## [1] ggplot.default* ggplot.function* ggplot.summaryP*
## [4] ggplot.transcan*
## see '?methods' for accessing help and source code
```

Although the `fortify()` function has a method for linear models that creates a data frame suitable for use with `ggplot()`, it extracts only the raw data, fitted values, and residuals. To show the effects of a model, it can be helpful to plot predicted values as a function of one variable holding other variables at specific values. The method that follows implements a system to do this.

It takes the same basic arguments as `ggplot()` but adds the `vars` argument, and each variable used in the model is passed with a specific set of values to hold it at for prediction. All possible combinations of these are created by passing the list as arguments to the `expand.grid()` function by using the `do.call()` function. The dependent variable, or `yvar`, is then extracted from the model formula and converted to a character. New predictions along with standard errors for the predictions are generated. A 95 percent confidence interval is generated with the lower limit, LL, and upper limit, UL, based on the fit or predicted value and the normal quantiles times the standard error of the fit. Finally, the column name is changed from fit to whatever the dependent variable's actual name was, and then the data for prediction and the predicted values are combined into a new data frame. This is passed to ggplot(), which dispatches to the ggplot() method for data frames:

```
ggplot.lm <- function(data, mapping, vars, ...) {
  newdat <- do.call(expand.grid, vars)
  yvar <- as.character(formula(data)[[2]])
  d <- as.data.frame(predict(data, newdata = newdat, se.fit = TRUE))
  d <- within(d, {
    LL <- fit + qnorm(.025) * se.fit
    UL <- fit + qnorm(.975) * se.fit
  })
  colnames(d)[1] <- yvar
  data <- cbind(newdat, d[, c(yvar, "LL", "UL")])
  ggplot(data = data, mapping = mapping, ...)
}
```

With this method in place, we can easily make some graphs from our linear regression model. The following code uses our new method, specifying the exact values to hold predictor variables, adds a line by using geom_line(), and uses a black-and-white theme with theme_bw(). The result is shown in Figure 5-2, and the code follows:

```
ggplot(m, aes(hp, mpg), vars = list(
                        hp = min(mtcars$hp):max(mtcars$hp),
                        vs = mean(mtcars$vs),
                        cyl = 8)) +
  geom_line(size=2) +
  theme_bw()
```

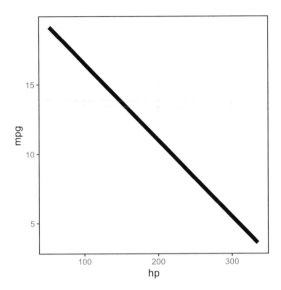

Figure 5-2. *Predicted regression line from the model, using the ggplot.lm() method*

Because all possible combinations of the predictor values are created by using expand.grid() in our method, we can make several predicted lines, such as holding vs at 0 and 1, shown in the following example code and in Figure 5-3.

```
ggplot(m, aes(hp, mpg, linetype = factor(vs), group = factor(vs)), vars =
list(
                    hp = min(mtcars$hp):max(mtcars$hp),
                    vs = c(0, 1),
                    cyl = 8)) +
  geom_ribbon(aes(ymin = LL, ymax = UL), alpha = .25) +
  geom_line(size=2) +
  theme_bw()
```

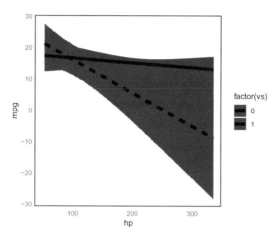

Figure 5-3. *Predicted regression lines from the model, with confidence intervals, using the ggplot.lm() method*

These examples provide a simple introduction to what can be done when writing new classes and methods or extending existing classes and methods. Because the S3 system is the most commonly used in R, it may be the single most important system to learn; it can be used so widely and can extend many classes of objects.

5.2 S4 System

In contrast with the S3 system, the S4 system is a formal system. The benefit of this formality is a greater assurance that objects of a particular class contain exactly what is expected. The downside of the formality is a greater complexity: more functions required to set up classes and methods and more rigid requirements in the programming. Whereas the S3 system allows for writing quick-and-dirty code, the S4 system requires more careful planning and comes with higher overhead.

Objects have three components: classes, slots, and methods. The name and structure of the object, what the object contains, is the class. The variables or other objects stored in the object are slots. Finally, the functions that can operate on an object are its methods. Throughout this section, we go over each of these components. For further reading on the S4 system, one excellent guide is *Software for Data Analysis: Programming with R* by John Chambers (Springer, 2010).

S4 Classes

Because S4 classes are more formal, planning is required before writing a new class. Unlike S3 classes, in S4 the names and types of every variable to be included must be specified. Previously we defined a class, textplot, by using the S3 system. We can define the same class by using the S4 system. The x and y arguments should both be numeric type, and the labels argument should be a character string. We also know that the length of all the arguments should be the same. Finally, we may want to assert that although it is okay to have a missing label, a blank label is not allowed. All of this needs to be considered and specified in advance because although in the S3 system none of this could be controlled, in the S4 system we can define everything when we create the class. To create our first S4 class, we use the setClass() function, shown in the next example. Not every argument is required, but it is considered a good practice to be as explicit as possible when defining a new class.

Note   S4 classes are created by calling setClass(). The Class and slots arguments are required. The class name is specified as a character string, and slots (holding variables) are defined as a named character vector; names correspond to slot names, and values correspond to the class of each slot. An "empty" object can be specified by using the prototype argument, and validity checking can be set by passing a function to the validity argument.

The first argument, Class, provides the name of the class. Next, the slots are defined by using a named vector in which the names indicate the slot names and the values indicate the type of each slot. The prototype argument is not required, but it is helpful to define, as it determines how to create an "empty" object or how an object of that class is created when no specific data is specified. The validity argument is also not required, but allows explicit checks to be run that the object conforms to expectations. R by default ensures the appropriate type of objects is passed to the slots, but many other tests and validity checks can be added to reduce the chances of an object being created that does not work as intended. Here is the setClass() example:

```
#### S4 System - Classes ####

setClass(
  Class = "textplot",
  slots = c(
    x = "numeric",
    y = "numeric",
    labels = "character"),
  prototype = list(
    x = numeric(0),
    y = numeric(0),
    labels = character(0)),
  validity = function(object) {
    stopifnot(
      length(object@x) == length(object@y),
      length(object@x) == length(object@labels))
    if (!all(nchar(object@labels) > 0, na.rm = TRUE)) {
      stop("All labels must be missing or non zero length characters")
    }
    return(TRUE)
  }
)
```

To create new objects of a particular class in the S4 system, we use the function new(). In the code that follows, we examine three attempts to create a new object. First, we make a correct and valid object. Then we look at what happens if we try to use the wrong type of argument. Finally, we look at an example that tests the validity function we created:

```
new("textplot",
    x = c(1, 3, 5),
    y = c(1, 2, 4),
    labels = c("First", "Second", "Third"))

## An object of class "textplot"
## Slot "x":
## [1] 1 3 5
```

```
##
## Slot "y":
## [1] 1 2 4
##
## Slot "labels":
## [1] "First"  "Second"  "Third"

new("textplot",
    x = c(1, 3, 5),
    y = c(1, 2, 4),
    labels = 1:3)
```

```
## Error in validObject(.Object): invalid class "textplot" object: invalid
    object for slot "labels"
```

```
new("textplot",
    x = c(1, 3, 5),
    y = c(1, 2, 4),
    labels = c("First", "Second", ""))
```

```
## Error in validityMethod(object): All labels must be missing or non zero
    length characters
```

These errors would not happen if we were using the S3 system to define a class, as no such type checking nor validity checking occurs. In the previous examples, each attempt to create a new object had at most one error. The next example has two errors: the vector passed to the y slot is not the same length as the other two, and there is a zero-length character label. However, as currently written, only the first error is caught, because the validity function stops as soon as there are any problems:

```
## multiple errors
new("textplot",
    x = c(1, 3, 5),
    y = c(1, 2),
    labels = c("First", "Second", ""))
```

```
## Error in validityMethod(object): length(object@x) == length(object@y) is
    not TRUE
```

Another way to write the validity function is so that, rather than throwing errors for any problems, the function collects them and returns all at the end. While revising the validity function, we could also think about how to make the error messages more informative. As it stands, it is fairly straightforward to see what the problem is, but perhaps not to see exactly what the problem is. For example, it is evident that the lengths are not equal, but is it that x is too long or y is too short? This is implemented in the revised code to define a class.

However, before diving into improving the validity function, we need a brief diversion on creating and formatting character strings in R. First, when writing text, new lines can be inserted by including the special character \n. To see the examples, we use the cat() function, which stands for concatenate, and print, and writes text out to the R console (or other locations if a file is specified). The following two examples are identical except for the line break between a and b in the second example:

```
## diversion on formatting text strings
cat("ab", fill = TRUE)

## ab

cat("a\nb", fill = TRUE)

## a
## b
```

One commonly used function is paste(), which can combine vectors or collapse them. Combining is shown in the first example that follows. The two vectors are combined by using the separator, defined by the sep argument, an empty string in our case. In the second example, a single vector with multiple elements is collapsed into a single character string. How the elements of the vectors are combined into one string is determined by the argument to collapse which is, in our example, the line break character:

```
paste(c("a", "b"), c(1, 2), sep = "")

## [1] "a1" "b2"

paste(c("a", "b"), collapse = "\n")

## [1] "a\nb"
```

These are useful functions for us when writing a validity-checking function, as they allow us to combine multiple errors into one string with line breaks as needed. The other key function we use is `sprintf()`. Its first argument is a user-defined string, with special symbols that always start with the percentage sign (%) where values should be substituted. The subsequent arguments are the values to substitute. An example may be the clearest way to show it. Here, %d is used to indicate that an integer is substituted, and then R substitutes in 98, 80, and 75. The order of substitution is the order of appearance:

```
sprintf("First (%d), Second (%d), Third (%d)", 98, 80, 75)

## [1] "First (98), Second (80), Third (75)"
```

Commonly used format options for substitutions are %d for integers, %f for fixed-point decimals, %s for strings, and % for a literal percentage sign. Each is demonstrated in this next example, and further documentation is available in the help pages, `?sprintf`. For the numeric value, we use 0.2 to specify that the number should be rounded to two decimal places:

```
sprintf("Integer %d, Numeric %0.2f, String %s, They won by 58%%", 5,
3.141593, "some text")

## [1] "Integer 5, Numeric 3.14, String some text, They won by 58%"
```

Armed with `paste()` and `sprintf()`, we can proceed to revise the validity function for our `textplot` class to provide more informative errors, and to run all checks, collecting errors along the way and returning all of them at the end. One final note: Previously, we used the idiom `new("classname", arguments)` to create a new object of a particular class. While this is perfectly acceptable, there is a shortcut. The function `setClass()` is primarily called for its side effect of defining a new class, but it also invisibly returns a constructor function. If we save the results from our call to `setClass()`, by convention in an object with the same name as the class, we can use the resulting object to create a new object of that class:

```
## Revised to catch multiple errors and be more informative
textplot <- setClass(
  Class = "textplot",
  slots = c(
    x = "numeric",
    y = "numeric",
```

```r
    labels = "character"),
  prototype = list(
    x = numeric(0),
    y = numeric(0),
    labels = character(0)),
  validity = function(object) {
    errors <- character()
    if (length(object@x) != length(object@y)) {
      errors <- c(errors,
                  sprintf("x (length %d) and y (length %d) are not equal",
                          length(object@x), length(object@y)))
    }
    if (length(object@x) != length(object@labels)) {
      errors <- c(errors,
                  sprintf("x (length %d) and labels (length %d) are not equal",
                          length(object@x), length(object@labels)))
    }
    if (!all(nchar(object@labels) > 0, na.rm = TRUE)) {
      errors <- c(errors, sprintf(
        "%d label(s) are zero length. All labels must be missing or non
        zero length",
        sum(nchar(object@labels) == 0, na.rm = TRUE)))
    }

    if (length(errors)) {
      stop(paste(c("\n", errors), collapse = "\n"))
    } else {
      return(TRUE)
    }
  }
)
```

Now when we create the same object with multiple problems, we get far more information and save some keystrokes. We can see the lengths of x and y and also learn that there are problems with the labels:

```
## multiple errors more informative
textplot(
  x = c(1, 3, 5),
  y = c(1, 2),
  labels = c("First", "Second", ""))
```

```
## Error in validityMethod(object):
##
## x (length 3) and y (length 2) are not equal
## 1 label(s) are zero length. All labels must be missing or non zero length
```

S4 Class Inheritance

So far, you have seen how to define new S4 classes. It may still seem like using the S4 system requires much more work than the S3 system, and with little benefit, aside from more formal validation and error checking. One of the powerful features of the S4 system is inheritance.

Note   S4 classes can inherit from existing S4 classes by calling `setClass()` with the additional argument contains = "S4 Class to Inherit". Existing slots, prototype, and validity checking are inherited. Only new slots and corresponding prototypes and validity checking need to be specified in the new `setClass()` call.

We previously created a simple `textplot` class. Now suppose that although sometimes our simple class is sufficient, at other times we may need more. For example, at times we may want to drill down into the data and create a panel of plots for different subsets of the data by some grouping variable. To this end, we want a `groupedtextplot` class. However, the only additional data we need is one more slot. One option is to copy and paste our old code and then modify it as needed. In this section, we explore how using inheritance lets us reuse and extend existing classes. When a class inherits from another class, all of the slots from the previous class are also inherited, as is validity checking. Next, we create our `groupedtextplot` class. It is similar to creating a new class, but we define only new slots and then specify the inheritance by using the argument contains. Because the validity checking is also inherited, we need to specify only validity checks for the new slots:

```r
# Revised to catch multiple errors and be more informative
groupedtextplot <- setClass(
  Class = "groupedtextplot",
  slots = c(
    group = "factor"),
  prototype = list(
    group = factor()),
  contains = "textplot",
  validity = function(object) {
    if (length(object@x) != length(object@group)) {
      stop(sprintf("x (length %d) and group (length %d) are not equal",
                   length(object@x), length(object@group)))
    }
    return(TRUE)
  }
)
```

With that small amount of code, we are ready to use our new class. In the following two examples, the first shows a correctly created new object, and the latter shows the familiar error messages when we attempt to create an invalid object:

```r
gdat <- groupedtextplot(
    group = factor(c(1, 1, 1, 1, 2, 2, 2, 2)),
    x = 1:8,
    y = c(1, 3, 4, 2, 6, 8, 7, 10),
    labels = letters[1:8])
gdat

## An object of class "groupedtextplot"
## Slot "group":
## [1] 1 1 1 1 2 2 2 2
## Levels: 1 2
##
## Slot "x":
## [1] 1 2 3 4 5 6 7 8
##
## Slot "y":
```

```
## [1] 1 3 4 2 6 8 7 10
##
## Slot "labels":
## [1] "a" "b" "c" "d" "e" "f" "g" "h"
```

```
groupedtextplot(
    group = factor(c(1, 1, 1, 1, 2, 2, 2, 2)),
    x = 1:8,
    y = c(1, 3, 4, 2, 6, 8, 7),
    labels = c(letters[1:7], ""))
```

```
## Error in validityMethod(as(object, superClass)):
##
## x (length 8) and y (length 7) are not equal
## 1 label(s) are zero length. All labels must be missing or non zero length
```

In this case, `textplot` would be called the parent class, and `groupedtextplot` would be called the child class. This relationship can be diagrammed (and sometimes for complex inheritance, diagramming is helpful). By convention, the relationship is shown graphically with an arrow pointing from the child to the parent(s), such as `textplot <- groupedtextplot`. Also by convention, parents are typically on the left, or above if graphing from top to bottom. Although we cover inheritance from only a single parent in this book, classes can inherit from multiple parents, and those parents can inherit from parents, and so on. It is in these cases where a visual diagram is particularly helpful. Another benefit of using inheritance, rather than writing a whole new class, is that methods are also inherited. This means that we can reuse both the slots from the parent and the methods written for the parent class, a topic we turn to in the next section.

S4 Methods

In addition to the `methods()` function you saw earlier to display the methods available for a given function, in the S4 system we can find methods by using `showMethods()`. Because of the more formal class system, `showMethods()` can be used to show all methods for a specific function or to show all methods (for any function) for a particular class by using the `classes = "class name"` argument. This can be helpful if you are working with a new class and want to know what methods have already been written and are available.

Note   Available S4 methods can be examined by using `showMethods("generic function")`. New S4 methods are defined by calling `setMethod()`. The main three arguments are f, signature, and definition, containing the name of the generic function (string), the S4 class name that will dispatch to this method (string), and a function that is the actual method, respectively.

To write new methods in the S3 system, we simply write functions with a special naming convention. To define S4 methods, we use the function `setMethod()`. It is possible to write new methods for existing classes, as well as writing methods for new classes, of course. For a new class, a method is needed for `show()`. When an object is simply typed at the console, R shows it by calling `show()`. Without a `show()` method for a new class, the default printing is quite ugly, as you have seen in our example so far.

In the following code, we define a new method for our textplot class. The first argument is the function name for which we create a method. The next argument, the signature, is the name of the class. In this case, because `show()` takes a single argument, only one class name needs to be specified. For functions with multiple arguments, the signature can become more complex, with different methods depending on the class of multiple arguments. Finally, we write our function, the definition of the method. The code is relatively simple, using the `cat()` function to display the values. The argument `fill = TRUE` has the effect of adding a line feed so that each line starts with X: or the variable label and then the values. The `head()` function is used to get at most the first five values, or less if there are fewer values:

```r
## method
setMethod(
  f = "show",
  signature = "textplot",
  definition = function(object) {
    cat("    X: ")
    cat(head(object@x, 5), fill = TRUE)
    cat("    Y: ")
    cat(head(object@y, 5), fill = TRUE)
    cat("Labels: ")
    cat(head(object@labels, 5), fill = TRUE)
})
```

R echoes the name of the generic function, show(), for which we just created a method. Now we get some nicer output when we create a textplot class object:

```
## nicer way of showing data
dat <- textplot(
  x = 1:4,
  y = c(1, 3, 5, 2),
  labels = letters[1:4])
dat
```

```
##        X: 1 2 3 4
##        Y: 1 3 5 2
## Labels: a b c d
```

Once we start defining methods, the benefits of class inheritance become even greater. Because groupedtextplot inherits from textplot, if no method is defined for groupedtextplot, it falls back to the method for textplot, if available. Although not perfect, because the grouping is not shown, this is still nicer than the default:

```
gdat
```

```
##        X: 1 2 3 4 5
##        Y: 1 3 4 2 6
## Labels: a b c d e
```

Next, we define a method for the [function, which is used to subset data. This function is more complex, as we must build a logic tree allowing several possible combinations of arguments. The first argument of the function, x, is for the object to be subset. By convention, i refers to rows or observations and j to columns or variables. If only i is not missing (i.e., rows or observations are specified), typically all variables are included. If only j (variables) is specified, typically all observations are included. If both are specified, only select variables and select observations are included. This is accomplished by using a series of if and else if statements. There are three new functions:

- validObject() is a generic function that executes the validity check, if present, for a specific object class.

- slotNames() returns a character vector giving the names of each slot.

- slot() works similarly to the @ operator, but can use character strings to extract slots by name.

```r
setMethod(
  f = "[",
  signature = "textplot",
  definition = function(x, i, j, drop) {
    if (missing(i) & missing(j)) {
      out <- x
      validObject(out)
    } else if (!missing(i) & missing(j)) {
      out <- textplot(
        x = x@x[i],
        y = x@y[i],
        labels = x@labels[i])
      validObject(out)
    } else if (!missing(j)) {
      if (missing(i)) {
        i <- seq_along(x@x)
      }

      if (is.character(j)) {
        out <- lapply(j, function(n) {
          slot(x, n)[i]
        })
        names(out) <- j
      } else if (is.numeric(j)) {
        n <- slotNames(x)
        out <- lapply(j, function(k) {
          slot(x, n[j])[i]
        })
        names(out) <- n[j]
      } else {
        stop("j is not a valid type")
      }
    }

    return(out)
  })
```

Now that the method is set, several examples of its uses are shown:

```
dat[]
```

```
##        X: 1 2 3 4
##        Y: 1 3 5 2
## Labels: a b c d
```

```
dat[i = 1:2]
```

```
##        X: 1 2
##        Y: 1 3
## Labels: a b
```

```
dat[j = 1]
```

```
## $x
## [1] 1 2 3 4
```

```
dat[j = "y"]
```

```
## $y
## [1] 1 3 5 2
```

```
dat[i = 1:2, j = c("x", "y")]
```

```
## $x
## [1] 1 2
##
## $y
## [1] 1 3
```

With a show and subsetting method defined for our textplot class, we can easily leverage those to make a show method for our groupedtextplot class by looping through the object by group, subsetting, and showing each subset:

```
setMethod(
  f = "show",
  signature = "groupedtextplot",
  definition = function(object) {
    n <- unique(object@group)
```

```
    i <- lapply(n, function(index) {
      cat("Group: ", index, fill = TRUE)
      show(object[which(object@group == index)])
    })
  })
```

gdat

```
## Group:  1
##      X: 1 2 3 4
##      Y: 1 3 4 2
## Labels: a b c d
## Group:  2
##      X: 5 6 7 8
##      Y: 6 8 7 10
## Labels: e f g h
```

5.3 Summary

This chapter has introduced the S3 and S4 systems in R for developing new classes and methods. The S4 system, in particular, can be complicated, with inheritance from multiple parent classes and methods that are specialized to the class of more than one argument. Even if you use and develop in R, you may only rarely develop new classes, as there are already classes for most data types. However, it can often be helpful to develop or extend existing methods, and at the very least, understanding how classes and methods work makes it easier to use existing ones.

Hopefully, this chapter is enough to get you started using these systems and to see their capabilities. Table 5-1 describes key functions covered in this chapter. In Chapter 6, we bundle functions and classes and methods together to make our own R package. The focus is on making an R package, so you do not need to be too comfortable with classes and methods. If your work would benefit from the use of the S4 system, you can get slightly more in-depth coverage from "How S4 Methods Work" by John Chambers (available online at http://developer.r-project.org/howMethodsWork.pdf). If you intend to work with the S4 system and need an in-depth dive, we recommend *Software for Data Analysis: Programming with R*. Another good book more focused on general R programming and the older S3 system is *S Programming*.

Table 5-1. *Listing of key functions described in this chapter and summary of what they do*

Function	What It Does
class()	Returns the class of an object (S3 or S4).
inherits()	Checks whether an object is of a certain class or inherits from that class.
methods()	Returns a list of the available methods for a given function.
showMethods()	Returns a list of the available methods for a given function or, if using the classes argument, available methods for any function for a given class.
:::	Operator that allows access to non-exported (nonpublic) functions from a package.
function. class()	Generic scheme for naming an S3 method, using the function name, followed by a period, followed by the class of object it should be applied to.
setClass()	Defines a new S4 class.
setMethod()	Defines a method for a particular function for a particular S4 class.
@	Low-level way to access slots by name in objects in the S4 system, similar to $ for other R objects or S3 class objects.
new()	Creates a new S4 object of a specific class.
paste()	Pastes strings together. Can operate on several vectors or collapse together all the elements of a single vector.
sprintf()	Formats strings and allows for substituting numbers and other strings into a defined template. Useful for making informative error messages or other messages to users.
show()	Generic function to show an R object. Also, the default function that is called when you type an object name at the R console.
'['()	Operator/generic function used to subset data or to access specific variables or rows. In the S4 system for a new class, methods must be defined, or the function is not usable.
head()	Lists the first few elements or rows of an object.

(continued)

Table 5-1. (*continued*)

Function	What It Does
validObject()	Generic function that checks when an object with an S4 class is valid by using the validity checks specified when the class was created (if any). Called by default when an object is created, but can also be called explicitly after modifying an object to check whether it is still valid.
slotNames()	Returns all the slot names of an S4 class object, similar to names() for other R objects or S3 class objects.
slot()	Can be used to access a specific slot of an S4 object.

CHAPTER 6

Writing Packages

Packages are the fundamental way to document, share, and distribute R code and data. Our goal is to write our own package. As a fair warning, this chapter is particularly complex because of the many software tools that are employed during R package development.

In this chapter, we make use of several packages. The devtools package [32] provides functions to help set up, document, and manage the development of an R package. The devtools package in turn installs and loads a whole set of other useful packages for development. The roxygen2 package [28] greatly eases writing documentation for R packages by allowing the function documentation to be written inline next to functions by using comments. The testthat package [24] is not required but has functions that facilitate quality control by testing that functions return expected values. This is good practice when writing a package to help make sure your code works as intended and to quickly identify any bugs that may get introduced later due to changes in R, other packages you use, or your own changes. The usethis package helps automate that testing (amongst other things) [43].

Similarly, the covr package [11] facilitates quality control by testing the percentage of package code that is executed (covered) by tests. Ideally, 100 percent of code would be covered by tests, though in practice, many R packages have no tests, so any coverage is better than average. In the package, we use one of the methods from Chapter 5 for ggplot2, so we also include the now familiar ggplot2 [25] package.

Finally, to help share and document the package, we use the pkgdown package [31]. This package allows us to create a web page automatically based off of the package documentation and files as well as any vignettes. Because GitHub can host web pages, if we host our package in a public GitHub repository, not only is the code readily available with version control, using pkgdown we can get a nice web page for it.

© Matt Wiley and Joshua F. Wiley 2020
M. Wiley and J. F. Wiley, *Advanced R 4 Data Programming and the Cloud*,
https://doi.org/10.1007/978-1-4842-5973-3_6

Writing packages can be a complex process. It may be helpful to familiarize yourself with the final product, available online at `https://github.com/JWiley/AdvancedRPkg`. You can also see the automatically generated website for the package at `http://joshuawiley.com/AdvancedRPkg`. We also recommend referring to our official GitHub record of the package as you go through this chapter, if you are unsure whether a file is in the correct location or set up correctly:

```
options(width = 70, digits = 2)
```

```
library(testthat)
library(devtools)
library(covr)
library(pkgdown)
library(ggplot2)
```

6.1 Before You Get Started

This chapter covers some of the basics of writing an R package and also introduces tools to facilitate the processes not included in the official manual, "Writing R Extensions," available online at `https://cran.r-project.org/doc/manuals/R-exts.html`. The manual can be quite technical, but it is the definitive guide. It is required reading if you plan to submit R packages to CRAN and a good idea to read even if you just plan to develop packages for your own use or a more limited user base such as your company or lab group.

Many R packages contain code from other programming languages such as C, C++, Fortran, or Java. Often this is because other languages, such as C, can be compiled and optimized to run much faster than R, so package developers may choose to write computationally intensive parts of their code in a compiled and highly efficient language. However, for this chapter, we discuss how to write only R packages containing pure R code or data, as the process is nearly identical, with most differences being idiosyncratic to the language included (e.g., makefiles).

Before developing R packages, some tools are required. Indeed, for those not used to software development, writing R packages can require a rather daunting toolchain. If you are using a Linux system, chances are you have many of these already or can readily install them. In Windows and Mac OS, it can take a bit more setup. Regardless of the system, the tools need to be accessible, and this means adding them to the system path so that when called from a command line, they can be found.

If using Windows, the main tools required (some command-line tools as well as compilers) are available from a binary installer at `https://cran.r-project.org/bin/windows/Rtools/`. Before the final Next of the install, there are two check boxes, one of which may be unchecked and adds this to your system path. Just select both.

Further details on the required Windows toolchain are available from the R manual at `https://cran.r-project.org/doc/manuals/R-admin.html#The-Windows-toolset`.

Also, you need LaTeX, a typesetting system. MiKTeX [21] is a popular choice (`www.miktex.org/`); be sure to select the 64-bit option if your system is 64 bit. While the default options are enough in most situations, we suggest MiKTeX users change to Yes the option for installing default packages on the fly.

If using Mac OS, get Xcode, developer tools, from the App Store. You may also need to get XQuartz (`www.xquartz.org/`). Additional details are available from the R manual (`https://cran.r-project.org/doc/manuals/R-admin.html#macOS-packages`). As on Windows, you need LaTeX. MacTeX is one choice (`www.tug.org/mactex/`).

Another option, instead of either MiKTeX or MacTeX, is `tinytex` [40]. The instructions for installing or running the package follow:

```
#install.packages("tinytex")
#tinytex::install_tinytex()
library(tinytex)
```

Two more decisions need to be made to progress to writing our first R package. First, what to name the R package? This seemingly simple task is complicated by the fact that over 15,000 R packages are currently on CRAN and the number is growing rapidly (`https://cloud.r-project.org/web/packages/available_packages_by_name.html`). Although you can give a package that is not to be uploaded to CRAN the same name as one on CRAN, doing so is problematic if you ever use the CRAN package or any other package that depends on it. Even if your package is not to be submitted to CRAN, someone else in the future might write a package with the same name and submit it to CRAN. For these reasons, choosing the package name requires some thought. Throughout this chapter, all the examples are to build one R package, called

AdvancedRPkg. The second decision is what license to use for the package. If you plan to submit to CRAN, the license must be compatible with CRAN. Even if the package is never sent to CRAN, a license may still be relevant. For this chapter, we use the GPLv3 license.

Version Control

Up until now, we have discussed relatively casual code development, some basic programming, functions, and classes. Developing new R packages is more complex, with many more files to manage. With this greater complexity, it is critical to have an efficient system for backing up files and for going back to earlier versions. Version control systems provide an excellent way both to back up code and to roll back changes to a prior version (e.g., if changes to implement a new feature break an old feature or introduce a bug). Various version control systems exist, but perhaps the most popular now is the open source Git originally developed to provide version control for the Linux kernel. Git can be used directly from the command line or through a graphical interface, of which there are several. Many people use GitHub (`https://github.com/`), a service that uses Git and hosts repositories online, freely for public projects. GitHub also provides a graphical interface for Windows and Mac OS (`https://desktop.github.com/`). However, if you do not want to learn or use Git or another version control system, feel free to ignore this section as well as any commands related only to version control throughout this chapter.

We use Git (and GitHub) for Windows in this chapter to provide version control for the example R package we develop. Specifically, this book uses the Windows GitHub desktop client. Git repositories can be used solely on a local computer, but using them with GitHub makes it easy to collaborate on packages and share early package code with others.

Setting up a Git repository is possible from the command line, but perhaps the most intuitive way for new users to create a new Git repository is online, directly through GitHub. We make a new repository for the R package we develop in this chapter and call it AdvancedRPkg. The repository is initialized with a README file, which we'll edit shortly. We can tell Git to ignore certain files or files with particular extensions by adding them to a file called .gitignore. Each file or extension is listed on its own line. We can edit it later, but for now, we just have GitHub create one. Figure 6-1 shows the steps to do this.

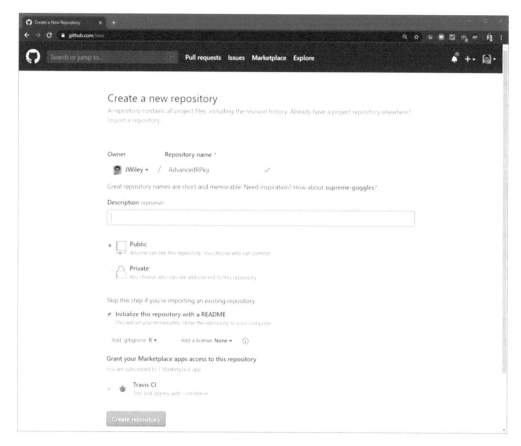

Figure 6-1. *GitHub page for creating the AdvancedRPkg repository*

If all works as planned, the results should look something like Figure 6-2. The repository is empty, except for the README and `.gitignore` files.

From here, we can clone the repository to our local computer. Essentially, this creates a local copy, where we do our package writing. To do this, open the GitHub desktop client, click "File," select "Clone repository," and then pick the repository, as shown in Figure 6-3.

We can pick the directory into which the repository should be cloned. Note that whatever directory you choose, a new directory is created for the repository with the same name as the repository (i.e., AdvancedRPkg). Settings for the repository can be managed via the terminal or by clicking on your repository and then going to settings in the GitHub desktop client. One of them shows the currently ignored files. Files are ignored based on the `.gitignore` file, and because we selected R earlier when creating the repository, some default file extensions that are often desirable to ignore have been included by default (Figure 6-4).

Windows often creates temporary (sometimes hidden) files, like `desktop.ini`, in directories. To avoid including these, we can edit the repository settings by adding a new line at the end and typing desktop.ini and saving. Then these files are ignored. Ignored files are those that the Git repository does not track and store changes in over time. They continue to exist in the directory; Git simply does not monitor and version them.

Aside from setting up a new Git repository, the tasks you are likely to do with Git are to commit changes to a repository and sync your local copy of the repository with the online (GitHub) version. Commits can be thought of as taking a snapshot of the state of files at the time of the commit. To avoid redundancy commits, snapshot only the changes made to files since the previous commit. Although even a slightly modified binary file may be difficult to compare, plain-text files, like R code, are easy to compare. Git is extremely efficient at tracking changes in text files over time and allowing you to either see what changes were made or go back to specific points in the past. Access to this history of a project (repository) is particularly helpful if errors or bugs are introduced into the code, so it can be pinpointed exactly when the bug was introduced (e.g., how many results based on the code may be impacted?) and see previous working versions of code.

In slightly more complex use cases, Git can be effectively used to manage different versions of the code. For example, it is common to have a project and periodically release stable or production-ready versions of the code while at the same time continuing development. This is accomplished in Git by using different branches of the same repository and periodically merging some of the changes from one branch (say, the development branch) into the stable branch.

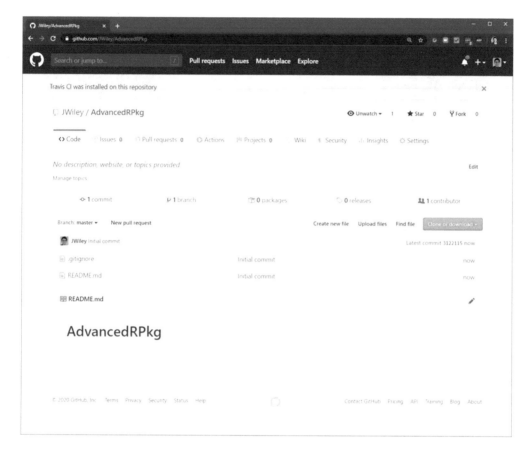

Figure 6-2. *GitHub page showing the created AdvancedRPkg repository*

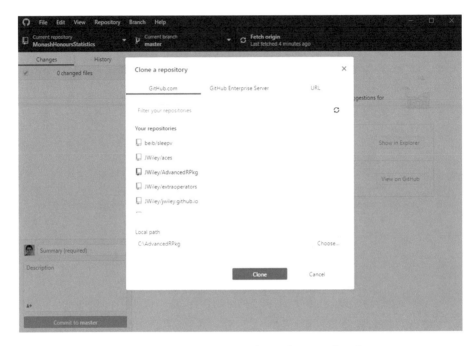

Figure 6-3. *GitHub desktop client cloning the AdvancedRPkg repository*

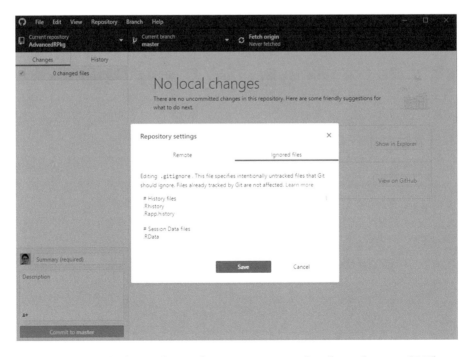

Figure 6-4. *GitHub desktop client showing settings for the AdvancedRPkg repository, including various file extensions that are ignored by the Git repository*

We show a few more relevant Git commands throughout this chapter. For more background on using Git, a great and free resource is *Pro Git* by Scott Chacon and Ben Straub (Apress, 2014), available at `https://git-scm.com/book/`. For questions and answers, Stack Overflow (`http://stackoverflow.com`) is a good resource. It is likely someone else has already asked a question similar to yours, and if not, you can ask and get answers. Also note that if you use RStudio for your R code, it has some integration with Git built in. For more on this, see RStudio Support at `https://support.rstudio.com/hc/en-us/articles/200532077-Version-Control-with-Git-and-SVN`. Finally, note you can access the online repository publicly at `https://github.com/JWiley/AdvancedRPkg`.

6.2 R Package Basics

An R package is just a directory with a particular set of files. Although the core of an R package is, of course, R code, many of the files are not strictly R code. Table 6-1 lists the primary files that may be in the root of an R package directory. Not all of these files are required, depending on other characteristics of the package. We'll get help creating some of these files by using the devtools package later in this chapter.

In addition to the root-level files, numerous subdirectories can be included in an R package. These are listed in Table 6-2, along with brief descriptions. For our relatively simple package, we work with only some of these, including R, man, tests, and docs.

Table 6-1. *Files that may be used in the root directory of an R package*

File	Description
DESCRIPTION*	Provides general information about the package. Required fields include Package (package name), Version, License, Title (package description), Author, and Maintainer (may be the same or different from the author). Also often includes information on dependencies, unless the package is stand-alone.
NAMESPACE*	Controls the package namespace, including which objects to export and which to import from other packages. Although required, we generate this automatically by using the roxygen2 package.

(continued)

Table 6-1. (*continued*)

File	Description
`README/README.md`	Not required and ignored by R, but helpful for readers and users. Provides general information on the package (where to get help, a brief overview, installation guidance, or whatever else would be useful in a brief document).
`NEWS`	Provides information on changes or news about the package. Commonly includes new features and bug fixes as well as any other major changes compared to previous versions.
`LICENSE`	A license file if one of the common licenses is not used or if additional information is required.
`INDEX`	Optional, as generated automatically from the documentation files. But if specified, provides a listing of all interesting/useful objects as a name and description on each line.
`configure, cleanup`	Optional Bourne shell scripts that are run before and after installation, respectively, on Unix systems (e.g., Linux, Unix).

*indicates a required file.

Starting a Package by Using DevTools

At a minimum, at the root directory of your package, you need `DESCRIPTION` and `NAMESPACE` files. You also need a subdirectory, R, which, sensibly, is where you put your R code. This is not enough for a package to submit to CRAN, but it is sufficient to start a functional package for private use. However, even for private use, proper documentation is incredibly helpful. The directory and file structure of packages is regrettably complex, especially to describe in text. Thus as a reference, Figure 6-5 is the final directory structure of our package that you should have by the end of this chapter, obtained from within the `AdvancedRPkg` directory.

For the initial setup, we use the function `create()`. We can specify more information, but the minimum is the path. We also set `open = FALSE` so the project is not automatically changed. To see what has been created, we can use the `list.files()` function, which shows that `DESCRIPTION` and `NAMESPACE` files and an R directory have been created. For the following code, either run it from the R command line or make a new R file (we called ours chapter06.R). It is important to adjust the path argument to point to the directory with the Git repository, wherever you cloned that on your local machine. In our case, R's working directory (which you can check by calling the getwd() function) is in the directory directly above the AdvancedRPkg directory. If your R session is not there, either adjust the path argument or change R's working directory to be in the parent directory to AdvancedRPkg by using the setwd() function:

```
create(
  path = "AdvancedRPkg/",
  open = FALSE)
```

```
## v Setting active project to 'C:/.../RCode/AdvancedRPkg'
## v Creating 'R/'
## v Writing 'DESCRIPTION'
## Package: AdvancedRPkg
## Title: What the Package Does (One Line, Title Case)
## Version: 0.0.0.9000
## Authors@R (parsed):
##     * First Last <first.last@example.com> [aut, cre] (<https://orcid.
        org/YOUR-ORCID-ID>)
```

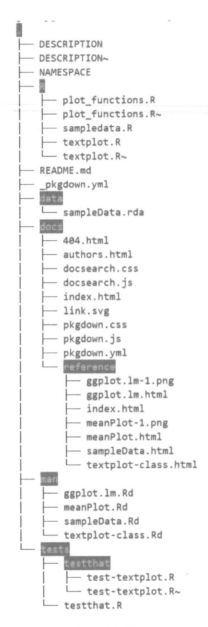

```
├── DESCRIPTION
├── DESCRIPTION~
├── NAMESPACE
├── R
│   ├── plot_functions.R
│   ├── plot_functions.R~
│   ├── sampledata.R
│   ├── textplot.R
│   └── textplot.R~
├── README.md
├── _pkgdown.yml
├── data
│   └── sampleData.rda
├── docs
│   ├── 404.html
│   ├── authors.html
│   ├── docsearch.css
│   ├── docsearch.js
│   ├── index.html
│   ├── link.svg
│   ├── pkgdown.css
│   ├── pkgdown.js
│   ├── pkgdown.yml
│   └── reference
│       ├── ggplot.lm-1.png
│       ├── ggplot.lm.html
│       ├── index.html
│       ├── meanPlot-1.png
│       ├── meanPlot.html
│       ├── sampleData.html
│       └── textplot-class.html
├── man
│   ├── ggplot.lm.Rd
│   ├── meanPlot.Rd
│   ├── sampleData.Rd
│   └── textplot-class.Rd
└── tests
    ├── testthat
    │   ├── test-textplot.R
    │   └── test-textplot.R~
    └── testthat.R

7 directories, 34 files
```

Figure 6-5. *Directory structure*

Table 6-2. *R package subdirectories*

Subdirectory	Description
R*	Directory that contains all the R source code for functions, classes, methods, and so forth.
man*	Directory that contains the R documentation files. We generate these automatically by using the roxygen2 package.
tests	Optional directory containing R code used to test that the package works as expected.
docs	Optional directory containing additional documentation, used by pkgdown for the package website.
data	Optional directory containing data to be shipped with the package. Typically, included data sets are small and used to illustrate a package's features, rather than as a primary means of sharing data.
demo	Optional directory containing demonstrations of the R package, as R source code files.
exec	Can contain additional required executable scripts. Only files are included, not subdirectories.
inst	Can contain additional files that are copied to the installation directory when the package is installed. For example, may be used to share a NEWS file with users or to include images or other nonstandard documentation.
po	Used to add translations of C and R error messages and other localization-related tasks.
src	Typically, source code from other languages, such as C, C++, or Fortran. This code is usually compiled; and if you use it, it often requires specifying a makefile.
tools	Not commonly used, but can be used to provide additional files required for configuration.
vignettes	Used to provide one or more vignettes or guides to using the package. Vignettes typically have more introductory background and complete examples of how you might use a package overall, compared with function documentation, which is generally unique to that function. Though not required, can be very helpful for users.

*indicates a required subdirectory.

```
## Description: What the package does (one paragraph).
## License: What license it uses
## Encoding: UTF-8
## LazyData: true
## v Writing 'NAMESPACE'
## v Setting active project to '<no active project>'

list.files("AdvancedRPkg")
[1] "DESCRIPTION" "NAMESPACE"    "R"                "README.md"
```

The output also shows some of the fields and the information used to fill them in. Since most of these are placeholders, we want to open the files by using a text editor (any should be fine, including RStudio) and modify them. From the directory, `create()` guesses the package name, but the rest need to be filled out. Using the editor, we change those fields to the following:

```
Package: AdvancedRPkg
Title: An Example R Package for the Book Advanced R
Version: 0.0.0.9000
Authors@R: c(
  person("Matt", "Wiley",
    email = "mattwiley@example.com", role = c("aut")),
  person("Joshua F.", "Wiley",
    email = "joshwiley@example.com", role = c("aut", "cre", "cph"))
  )
Description: This package will demonstrate the basics of an R
  package including documentation and tests.
Depends: R (>= 3.6.0)
License: GPL (>= 3)
Encoding: UTF-8
LazyData: true
```

Note the version number, which is designed in the format `Major.Minor.Patch.DevelopmentVersion`. A good overview of the considerations in determining software versions is described in *Semantic Versioning Specification* at `https://semver.org/`. It is quite prescriptive in its recommendations, but it is often helpful to have a fixed set of rules in place for determining a major or minor version of the software. The final piece, the DevelopmentVersion, is present only in development versions and is dropped for release. Despite these rules and guides, the reality of many R packages is far more heterogeneous and inconsistent (not everyone agrees on these rules, and even those who agree do not always strictly follow them).

Next up are the authors, described using the `person()` function. In addition to each individual's name and email, three-letter abbreviations are used to describe roles and relations. All possibilities are outlined in the MARC Code List for Relators at `www.loc.gov/marc/relators/relaterm.html`. However, the most commonly used ones are `aut` for the author, `cre` for a creator who is the person responsible for the project (or, in R terms, the person to complain to if there are problems, ergo a maintainer), and `cph` for the copyright holder. Then we provide a few sentences for a description of the package and specify the R version our package depends on, the license, and that data can be loaded on demand (saving startup time).

Adding R Code

With the basics of a package in place, it is time to start adding some R code, the whole point of a package! Since the focus of this chapter is on packaging the code, we reuse some functions and classes from previous chapters. As a first step, we create two files located in the `AdvancedRPkg/R/` directory: `plot_functions.R` and `textplot.R`.

Next, we copy the final `meanPlot()` function from Chapter 4 to make a plot with means. We add the code for this function along with the S3 method, `ggplot.lm()`, we wrote in Chapter 5 and put both in `plot_functions.R`. Then, we copy the classes and methods (show and subset, which is the bracket operator [ ]) for the textplot class from Chapter 5 into `textplot.R`. Because it is easy to copy the wrong code, we suggest copying and pasting the code from our GitHub repository (`https://github.com/JWiley/AdvancedRPkg`). If everything works, it should look like this:

```
list.files("AdvancedRPkg/R")
## [1] "plot_functions.R" "textplot.R"
```

As a side note, it can be difficult to decide how many separate files to have. It does not matter for R, but it does make a difference for development. It is not good to have all your code in one file nor to split every function, class, and method into a separate file. If the result is not too long, we try to group related functions (such as plotting functions) and to group related classes and methods. As common sense, even if you can, do not give files the same names differentiated only by use of lowercase or uppercase letters; it is also generally a bad idea to use special characters or symbols in filenames.

Now that we have a basic package template, we can load the functions by using the load_all() function from devtools. All we have to do is specify the path to the package. We can now see that some of our functions are available in the package namespace, which we do by using the ls() function and specifying where we want it to list available objects:

```
load_all("AdvancedRPkg")
## Loading AdvancedRPkg

ls(name = "package:AdvancedRPkg")
## [1] "ggplot.lm"          "library.dynam.unload"
## [3] "meanPlot"           "system.file"
## [5] "textplot"
```

In Chapter 4, we briefly discussed scoping. Scoping becomes more important to understand when writing packages. Package authors can write functions with the same names as functions in other packages. Although these functions do not overwrite each other, it can be confusing to be clear about which function you intend to call. This is where the package namespace can be helpful. The namespace controls which functions from other packages are imported (and therefore used by code from within that package) and which functions from a package are exported so that they are publicly available to users of the package. You can import specific functions from other packages, using the import feature. If you want many functions from another package, you may decide to make your package depend on that other package, in which case all public functions are available to your package. Another difference between importing and depending on another package is what happens when your package is loaded. If package B depends on package A when a user calls library(B), package B and package A are both loaded and attached. If package B imports from package A, when a user calls library(B), package B gets loaded and attached, and package A is loaded but not attached. Because package A is loaded, its functions are available to code within package B, but they are not exposed directly to the user because package A was not attached. Even when package A (or package B) is loaded and attached, only the exported functions are publicly available to users. This is beneficial, as it allows package developers to write and document functions that are for internal use only. If you do not need to export a function, it is a good idea not to. If two R packages export functions of the same name and a user loads and attaches both packages, the function from whichever package is loaded later masks the earlier one. For a user, the only choice then is either to not load and attach one package or to be explicit with the function calls, using the double colon operator, `PkgA::foo()`, `PkgB::foo()`. With thousands of R packages and even more functions, exporting only necessary functions helps avoid such conflicts and masking of names.

All of this is controlled via the `NAMESPACE` file of a package. Although we did not look at it, our package does have a `NAMESPACE` file, which was created when we ran `create()`. However, the `load_all()` function is unique in that by default when it loads a package, it exports all objects. This is because during development, it is often convenient to be able to call all functions whether they are exported or not.

Tests

With functions and methods added to our new package and working in R, we can begin to think about quality control. It is great to write code that runs, but it is also crucial that the code does what is intended. Now, what is intended and what is expected are not always the same (that is where documentation plays a critical role, which we cover next). Still, even for the developer, tests ensure that what is written does what it is supposed to do.

Tests may be written at any stage of package development. If functions, names, and features are radically changing, perhaps it is too early to start writing many tests. However, even if a package is not done, if some functions are relatively stable, writing tests early can be helpful. Writing tests earlier rather than later is helpful because it is often easier to write a test immediately after writing a function, when its purpose and the way it works are still fresh in your memory. (If you do not know or have forgotten how a function works, it is hard to test it adequately!) It is also helpful because if later functions build on earlier functions, testing along the way can help ensure that problems are due to the newly written functions and not to some previous building block.

Benefits aside, the practicalities of writing tests are made easier by the `testthat` package. To get tests set up, we use some functions from the `usethis` package, particularly the `use_test()` function. This function takes the name of a test and will create the relevant directories and subdirectories needed for the `testthat` package. We also change directories to be in our package directory first so the relative paths are all correct. The output shows a variety of tasks that the use_test() function has executed, including adding needed packages to our DESCRIPTION file and creating directories and skeleton R files for use to fill in with tests:

```
setwd("AdvancedRPkg")
use_test("textplot")

## v Setting active project to 'C:/.../RCode/AdvancedRPkg'
## v Adding 'testthat' to Suggests field in DESCRIPTION
## v Creating 'tests/testthat/'
## v Writing 'tests/testthat.R'
## * Call 'use_test()' to initialize a basic test file and open it for
   editing.
## v Increasing 'testthat' version to '>= 2.1.0' in DESCRIPTION
## v Writing 'tests/testthat/test-textplot.R'
## * Modify 'tests/testthat/test-textplot.R'
```

To have the tests run, we need to indicate that our package requires the testthat package. The core functionality of the package does not depend on testthat, so instead we add it to the DESCRIPTION file under a new section, Suggests. Normally, you would have to do this by hand, but as you can see from the earlier output, the usethis package did that for us.

It is rather difficult to test graphing functions properly, so for our tests, we check whether the subset method for textplot works as intended. In the file called test-textplot.R located under AdvancedRPkg/tests/testthat/, we set up a simple textplot class object and then run a series of tests. The tests have two components. First is an outer call to test_that(). The first argument is a description of the test, and the second is code to do the test. The code can consist of anything, but commonly includes calls to one of the expect_*() functions, of which there are many. We use the apropos() function call to show all the options for expect_*() before we make our choices:

```
apropos("expect_")
##  [1] "expect_condition"          "expect_cpp_tests_pass"
##  [3] "expect_equal"              "expect_equal_to_reference"
##  [5] "expect_equivalent"         "expect_error"
##  [7] "expect_failure"            "expect_false"
##  [9] "expect_gt"                 "expect_gte"
## [11] "expect_identical"          "expect_invisible"
```

```
## [13] "expect_is"              "expect_known_failure"
## [15] "expect_known_hash"       "expect_known_output"
## [17] "expect_known_value"      "expect_length"
## [19] "expect_less_than"        "expect_lt"
## [21] "expect_lte"              "expect_mapequal"
## [23] "expect_match"            "expect_message"
## [25] "expect_more_than"        "expect_named"
## [27] "expect_null"             "expect_output"
## [29] "expect_output_file"      "expect_reference"
## [31] "expect_s3_class"         "expect_s4_class"
## [33] "expect_setequal"         "expect_silent"
## [35] "expect_success"          "expect_that"
## [37] "expect_true"             "expect_type"
## [39] "expect_vector"           "expect_visible"
## [41] "expect_warning"
```

We choose expect_is() to check the class of the object and expect_equal() to check other features. It is also possible to check that your error checking is working, using expect_error() or expect_warning(), which "pass" if the code creates an error. We place the following code into our test-textplot.R file:

```
dat <- textplot(
  x = 1:4,
  y = c(1, 3, 5, 2),
  labels = letters[1:4])

test_that("textplot subset method works with rows only", {
  tmp <- dat[i = 1:2, ]
  expect_is(tmp, "textplot")
  expect_equal(length(tmp@x), 2)
})

test_that("textplot subset method works with variables only", {
  tmp <- dat[, j = c("x", "y")]
  expect_is(tmp, "list")
  expect_equal(length(tmp), 2)
  expect_equal(length(tmp$x), 4)
})
```

```
test_that("textplot subset method works with rows and variables", {
  tmp <- dat[1:2, j = c("x", "y")]
  expect_is(tmp, "list")
  expect_equal(length(tmp), 2)
  expect_equal(length(tmp$x), 2)
})
```

We can run the code tests directly. If everything works, there is no output. Output is created only when something does not pass the checks. However, typically, this code is not run directly; it is run in batches. We run all tests by using the devtools function, test(), and our package name/path. Since we changed directories so that the working directory *is* our package directory, we can leave it blank. The devtools package knows the directory structure for the testthat package, and so they play nicely together. To run the tests, test() first loads the package by using load_all() and then executes the tests. Back in the folder above the package directory, from our chapter-level R file, chapter06.R, we run all the tests by executing the code that follows:

```
test()
## Loading AdvancedRPkg
## Testing AdvancedRPkg
## v |  OK F W S | Context
## v |   8       | textplot
##
## == Results ====================================================
## OK:      8
## Failed:  0
## Warnings: 0
## Skipped:  0
```

Now, we can use the covr package to check code coverage. The covr package (https://covr.r-lib.org/) checks whether different parts of the code are run when a test is executed. For example, a function that contains if/else statements may have only part of its code executed by one test, unless additional tests are derived to check the other conditions. Again, although not required to develop a package, it is a helpful tool to see how well the current tests cover the package functionality. A high level of coverage (aiming toward 100 percent) is a good way to catch bugs and to ensure that new code development does not break old features and functionality; and all of this can also help to assure users that your package is a good choice.

Back in our top-level chapter06.R file, we run the package_coverage() function to test how well our tests cover our package's code. The only argument required is the directory where the package is located. In our case, R's working directory is the parent directory, AdvancedRPkg/. If the working directory for your R instance is different, you need to specify the appropriate path to get to the AdvancedRPkg/directory on your machine:

```
cov <- package_coverage()
```

The output (shown next) indicates current testing coverage is about 36 percent, and this is driven by coverage of code from textplot.R. It is also possible to get a more detailed analysis, using as.data.frame(cov). The output is substantial, so we show just a few rows and columns, but it is an invaluable resource if you think you have tests covering everything and need to figure out what aspects of your code are not being tested yet:

```
cov
## AdvancedRPkg Coverage: 33.33%
## R/plot_functions.R: 0.00%
## R/textplot.R: 51.61%

as.data.frame(cov)[1:3, c(1, 2, 3, 11)]
##                 filename functions first_line value
## 1 R/plot_functions.R   meanPlot          2     0
## 2 R/plot_functions.R   meanPlot          3     0
## 3 R/plot_functions.R   meanPlot          4     0
```

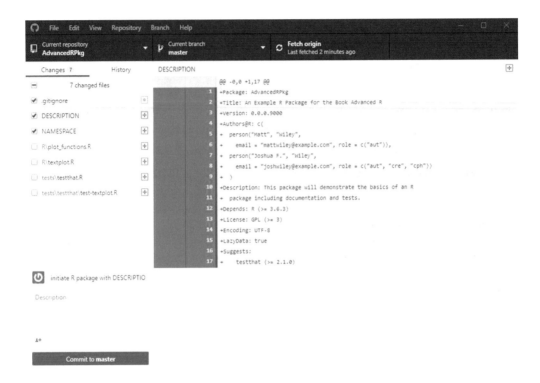

Figure 6-6. GitHub desktop client selecting changed files, adding a commit message, and committing to the master branch

Finally, with all the changes we have made to our package, it is time to check in with version control. It is possible to do this from the command line, but we use the GitHub desktop client. By default, all changed files are selected, but rather than commit all changes at once, we do them in related batches of files, along with meaningful messages. The commit messages are added along with each commit and serve to remind you or to let others know a high-level summary of what changes were made or why. It is not necessary to detail every change, as Git takes care of tracking exactly which files changed and how their contents changed.

To begin, we select the DESCRIPTION, NAMESPACE, and .gitignore files and add the message initiate R package with DESCRIPTION and NAMESPACE. The process is shown in Figure 6-6.

Next, we add plot_functions.R and textplot.R and add the commit message initial commit of R functions. Finally, we add testthat.R and test textplot.R and add the commit message adding testing for quality control. Of course, you also could have committed files along the way as they were created. If we switch over to the History tab of the GitHub desktop client, we can see the changes made. If we are happy for them to go public, we click the Sync button, which pulls changes from the GitHub repository to the local repository and pushes all committed changes from the local repository to the GitHub repository. Note that before you sync, all changes should be committed, because what get synced are not the actual files in the directory but the Git repositories. Changes to files get added to the local Git repository only when they are committed, so if you do not commit, the updates are not added to the local repository and so do not get pushed to the remote (GitHub) repository. Figure 6-7 shows what this looks like.

We have only scratched the surface of testing, and obviously our package has a long way to go before it is thoroughly tested. However, these tools should get you started on the right track (and ahead of the majority of the R packages on CRAN) regarding testing and quality control. The next crucial piece is the documentation, a topic we turn to next.

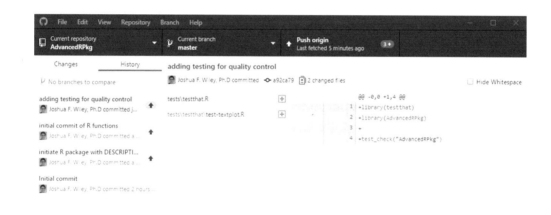

Figure 6-7. *GitHub desktop client showing the history of commits and the Sync button at the top right*

151

6.3 Documentation Using roxygen2

R documentation for functions, data, classes, and methods is located in the man subdirectory of a package and is done by using special R documentation files with the extension .Rd. However, for the programmer, this can be harder because the documentation is separated from the actual R code. There is also quite a bit of markup that must be written that is just part of the standard template. The roxygen2 package fully automates this process, including the subdirectory creation, and provides an easier way to write documentation by using specially formatted comments next to the R code. Running the roxygen2 package then converts these particular comments into appropriately formatted .Rd files in the man directory, so that R has everything it needs to create the help files for the package functions. In this section, we are going to look at how to document several types of objects including functions, data, classes, and methods.

Functions

To document objects using roxygen2, documentation is added after a special comment lead ##' or #'. Because # is R's comment character, R ignores it, but the special single quote indicates to roxygen2 that this information should be processed into the documentation. Note this syntax precludes commenting this code. All of this code is added to plot_functions.R above the meanPlot() function code.

As stated, comments are precluded because of syntax; thus, the order becomes vital. To see which part maps to which formal, just keep in mind there are three sections simply separated by line breaks: title, description, and details. The details section is optional. We add spaces between line breaks for reader clarity. Function arguments are indicated by using @param argument_name brief description of the argument. The argument names listed here must exactly match the named arguments in the function. The @return section is where we can indicate what sort of object is returned by the function. In this case, the function is called for the side effect of producing a plot, and the value it returns is not important. Any text can come after @author to indicate who wrote the function. Typically, the @author section is not required if the author is the same as the overall package author. For the function to be publicly available, it must be exported. This is accomplished by using the directive @export. The roxygen2 package translates this into additional lines of code in the NAMESPACE file, indicating it should be exported.

The @keywords section is optional. Finally, one of the most useful sections for your readers is the @examples section, which gives readers executable examples and is one of the easiest ways to show how to use a function:

```
##' Function to plot data and mean summary
##'
##' This is a simple function designed to facilitate plotting raw
##' data along with dots indicating the mean at each x-axis value.
##'
##' Although this function can be used with any type of data that works
##' with \code{plot}, it works best when the x-axis values are discrete,
##' so that there are several y-values at the same x-axis value so that
##' the mean of multiple values is taken.
##'
##' @param formula A formula specifying the variable to be used on the
##'    y-axis and the variable to be used on the x-axis.
##' @param d A data.frame class object containing the variables specified
##'    in the \code{formula}.
##' @return Called for the side effect of creating a plot.
##' @author Wiley
##' @export
##' @keywords plot
##' @examples
##' # example usage of meanPlot
##' meanPlot(mpg ~ factor(cyl), d = mtcars)
```

Now to make the documentation, we can run the roxygen2 package. This can be done directly by using the roxygenize() function and giving it the path to the directory of our package, or it can be done by using the document() function from the devtools package. We specify the package directory, and the rest happens automatically. The output shows that a documentation file was written (meanPlot.Rd) and the NAMESPACE file was written:

```
document()
## Updating AdvancedRPkg documentation
## Updating roxygen version in ~\AdvancedRPkg/DESCRIPTION
## Writing NAMESPACE
```

```
## Loading AdvancedRPkg
## Writing NAMESPACE
## Writing meanPlot.Rd
```

When built or installed, the R documentation file is converted to HTML and PDF. However, already you can preview the development version of the documentation. From the R console, the usual way of getting help for a function should now work:

```
?meanPlot
```

If you use RStudio, you can also open the file and preview it. The resulting HTML file (after building the package) is shown in Figure 6-8. Because of the simplicity of this function and also for the sake of space, we wrote fairly minimal documentation. Often, it is helpful for users and future reference to document more extensively and to carefully explain what each argument can and cannot take. If functions implement new procedures or statistical methods, it is also common to include some references by using the @references section.

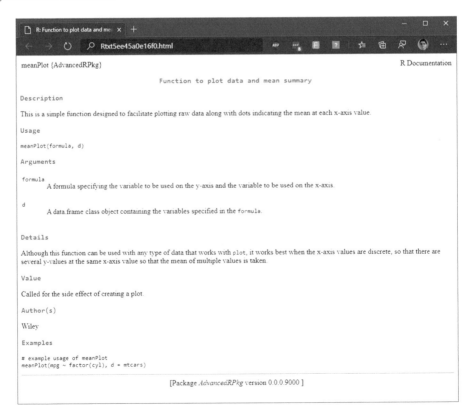

Figure 6-8. HTML output of the R documentation file for the meanPlot() function

Data

Before we can document data, we need to add some to our package. First, we make a small sample data frame, `sampleData`. Previously, we would make a data subdirectory in our package folder, located at `AdvancedRPkg/data/`, and then save the data as an `.rda` file. However, the `use_data()` function from the `usethis` package will take care of creating the correct directory and saving the data file in the correct format, in the correct place, with the correct name:

```
sampleData <- data.frame(
  Num = 1:10,
  Letter = LETTERS[1:10])

use_data(sampleData)
## v Setting active project to 'C:/.../RCode/AdvancedRPkg'
## v Creating 'data/'
## v Saving 'sampleData' to 'data/sampleData.rda'
```

We add the following code to document the data in a new file called sampledata.R located in the R subdirectory (i.e., `AdvancedRPkg/R/sampledata.R`). This provides a title and details on the data, along with the special parameters `@format` to indicate the format or type of data and `@source` to indicate where it is from. In quotes at the end, the name of the object is given:

```
##' Numbers and letters.
##'
##' A sample data set containing 10 numbers and letters with two variables:
##'
##' \itemize{
##'    \item Num. A number.
##'    \item Letter. An upper case letter (A to J)
##' }
##'
##' @format A data frame with 10 rows and 2 variables
##' @source Created as a sample
"sampleData"
```

After re-roxygenizing the package by using the document() function, the resulting HTML is shown in Figure 6-9, again by using the help utilities from the fully built package or by using preview from RStudio.

Note that only the new files are written when we document(). Our plot documentation has not changed, so there's no need to rewrite the file:

```
document()
## Updating AdvancedRPkg documentation
## Writing NAMESPACE
## Loading AdvancedRPkg
## Writing NAMESPACE
## Writing sampleData.Rd
```

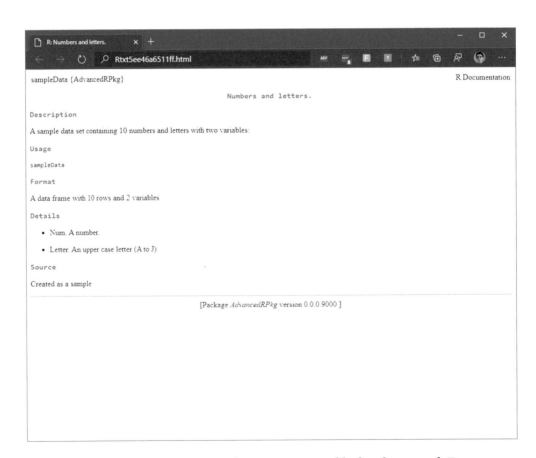

Figure 6-9. *HTML output of the R documentation file for the sampleData data set*

Classes

S3 classes are not formal and are not typically documented. S4 classes are straightforward to document. Their documentation goes immediately above the call to setClass() in our textplot.R file (AdvancedRPkg/R/textplot.R) with a title, a details section, and then some parameters—one @slot for each slot name, detailing the name and function of the slot. To use the S4 system in a package, we also need to add the methods package, which we do by adding @import methods and also adding it to the Imports field of the DESCRIPTION file. We can do this again using the usethis package, specifically the use_package() function specifying the package name and which section it belongs under:

```
use_package("methods", "Imports")
```

The full documentation for the class is shown in the following code:

```
##' An S4 class to hold text and Cartesian coordinates for plotting
##'
##' A class designed to hold the data required to create a textplot
##' where character strings are plotted based on x and y coordinates.
##'
##' @slot x A numeric value with the x axis coordinates.
##' @slot y A numeric value with the y axis coordinates.
##' @slot labels A character string with the text to be plotted
##' @import methods
```

After re-roxygenizing the package by using the document() function, the resulting HTML is shown in Figure 6-10.

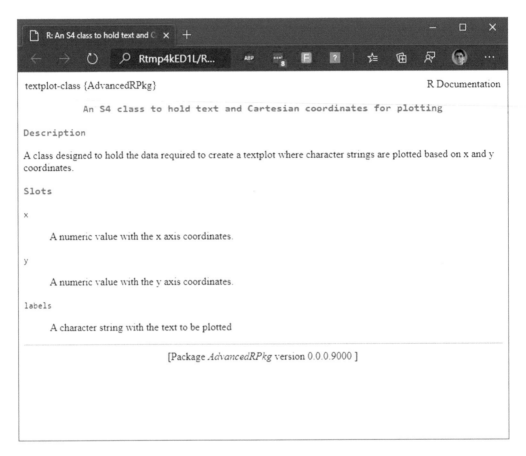

Figure 6-10. *HTML output of the R documentation file for the S4 class textplot*

Methods

Documenting methods is similar to documenting regular functions. S3 methods can be undocumented or documented as a function. The following is the roxygen2-style documentation added for the ggplot.lm method. The code is added immediately above the function definition in AdvancedRPkg/R/plot_functions.R.

Note that here we import the ggplot2 package so that it is available. We also add the ggplot2 package under the Imports section in the DESCRIPTION file, separated by a comma from the R version on which the package depends. We can do this again using the usethis package, specifically the use_package() function specifying the package name and which section it belongs under:

```
use_package("ggplot2", "Imports")
```

By importing it, roxygen2 will automatically add the appropriate import code to the NAMESPACE file:

```
##' Method for plotting linear models
##'
##' Simple method to plot a linear model using ggplot
##' along with 95% confidence intervals.
##'
##' @param data The linear model object from \code{lm}
##' @param mapping Regular mapping, see \code{ggplot} and \code{aes} for details.
##' @param vars A list of variable values used for prediction.
##' @param \ldots Additional arguments passed to \code{ggplot}
##' @return A ggplot class object.
##' @export
##' @import ggplot2
##' @examples
##' ggplot(
##'   lm(mpg ~ hp * qsec, data = mtcars),
##'   aes(hp, mpg, linetype = factor(qsec)),
##'   vars = list(
##'     hp = min(mtcars$hp):max(mtcars$hp),
##'     qsec = round(mean(mtcars$qsec) + c(-1, 1) * sd(mtcars$qsec)), 1)) +
##'   geom_ribbon(aes(ymin = LL, ymax = UL), alpha = .2) +
##'   geom_line() +
##'   theme_bw()
```

S4 methods can be documented alone, documented with the generic function, or documented with the class. This can be accomplished by using the @describeIn parameter, which is used in place of the title. For our S4 methods, we document them along with the class. The roxygen2 code is relatively brief. We could add details but do not need to. The roxygen2 package takes care of registering the methods, so all we really need to do is add any special notes and document use of parameters. The documentation for the show() method is as follows:

```
##' @describeIn textplot show method
##'
##' @param object The object to be shown
```

We follow this with the documentation for the [operator method. Note that we also export the [method and add an alias, which is just another way that users can look up the method:

```
##' @describeIn textplot extract method
##'
##' @param x the object to subset
##' @param i the rows to subset (optional)
##' @param j the columns to subset (optional)
##' @param drop should be missing
##' @export
##' @aliases [,textplot-method
```

After re-roxygenizing by using document(), the updated HTML help file is shown in Figure 6-11, including the class and additional methods documentation.

```
document()
## Updating AdvancedRPkg documentation
## Writing NAMESPACE
## Loading AdvancedRPkg
## Writing NAMESPACE
## Writing ggplot.lm.Rd
## Writing textplot-class.Rd
```

6.4 Building, Installing, and Distributing an R Package

Much as with Git, when building an R package, it can be helpful to ignore some of the files in the directory. We create a file called .Rbuildignore containing the following code in our main directory AdvancedRPkg/. This file should be located at the same level as the DESCRIPTION file:

```
desktop.ini
.Rhistory
.RData
```

After ensuring the documentation is up-to-date by running document() again, we are almost set. At this point, we have made some changes, so it would be a good idea to add and commit the changes to the Git repository (if you opted into that at the beginning). Although you could add all files at once by running git add, it is more informative and easier to revert later (if needed) if changes are committed in chunks based on similar topics. For example, we used separate commits for each of the following, which are shown in the code that follows in chunks:

Figure 6-11. *HTML output of the R documentation file for the S4 class textplot with added methods documentation*

- The updated code files (every file in the AdvancedRPkg/R/ directory), DESCRIPTION, and NAMESPACE, with the commit message added roxygen-style documentation

- The sample data, with the commit message added sample data

- The updated documentation files (every file in the AdvancedRPkg/ man/ directory), with commit message re-roxygenized package

- .Rbuildignore file, with the commit message adding file to ignore files during package build

Now we can build the package into a compressed source tar ball by using the build() function from the devtools package in our chapter06.R file. The output shows the build process and that a .tar.gz file is produced at the end:

```
build("AdvancedRPkg")
## v   checking for file 'C:\...\RCode\AdvancedRPkg/DESCRIPTION'
## -   preparing 'AdvancedRPkg':
## v   checking DESCRIPTION meta-information
## -   checking for LF line-endings in source and make files and shell
        scripts
## -   checking for empty or unneeded directories
## -   looking to see if a 'data/datalist' file should be added
## -   building 'AdvancedRPkg_0.0.0.9000.tar.gz'
##
## [1] "C:/.../RCode/AdvancedRPkg_0.0.0.9000.tar.gz"
```

We could install and use this. However, it can be helpful to run one more set of checks. The check() function in the devtools package runs Rcmd check on the package. The argument cran = TRUE uses the tests for CRAN. The following code should be run from our chapter06.R file:

```
check("AdvancedRPkg", cran = TRUE)
## Updating AdvancedRPkg documentation
## Writing NAMESPACE
## Loading AdvancedRPkg
## Writing NAMESPACE
## -- Building ---------------------------------------- AdvancedRPkg --
## Setting env vars:
```

```
## * CFLAGS : -Wall -pedantic
## * CXXFLAGS : -Wall -pedantic
## * CXX11FLAGS: -Wall -pedantic
## -------------------------------------------------------------------------
## v  checking for file 'C:\...\RCode\AdvancedRPkg/DESCRIPTION'
## -  preparing 'AdvancedRPkg':
## v  checking DESCRIPTION meta-information
## -  checking for LF line-endings in source and make files and shell
       scripts
## -  checking for empty or unneeded directories
## -  looking to see if a 'data/datalist' file should be added
## -  building 'AdvancedRPkg_0.0.0.9000.tar.gz'
##
## -- Checking ----------------------------------------- AdvancedRPkg --
## Setting env vars:
## * _R_CHECK_CRAN_INCOMING_USE_ASPELL_: TRUE
## * _R_CHECK_CRAN_INCOMING_REMOTE_  : FALSE
## * _R_CHECK_CRAN_INCOMING_  : FALSE
## * _R_CHECK_FORCE_SUGGESTS_  : FALSE
## * NOT_CRAN : true
## -- R CMD check ----------------------------------------------------------
## * using log directory 'C:/.../Temp/.../AdvancedRPkg.Rcheck'
## * using R version 3.6.3 (2020-02-29)
## * using platform: x86_64-w64-mingw32 (64-bit)
## * using session charset: ISO8859-1
## * using options '--no-manual --as-cran'
## * checking for file 'AdvancedRPkg/DESCRIPTION' ... OK
## * this is package 'AdvancedRPkg' version '0.0.0.9000'
## * package encoding: UTF-8
## * checking package namespace information ... OK
## * checking package dependencies ... OK
## * checking if this is a source package ... OK
## * checking if there is a namespace ... OK
## * checking for executable files ... OK
## * checking for hidden files and directories ... OK
```

```
## * checking for portable file names ... OK
## * checking whether package 'AdvancedRPkg' can be installed ... OK
## * checking installed package size ... OK
## * checking package directory ... OK
## * checking for future file timestamps ... OK
## * checking DESCRIPTION meta-information ... WARNING
## * checking top-level files ... OK
## * checking for left-over files ... OK
## * checking index information ... OK
## * checking package subdirectories ... OK
## * checking R files for non-ASCII characters ... OK
## * checking R files for syntax errors ... OK
## * checking whether the package can be loaded ... OK
## * checking whether the package can be loaded with stated dependencies
##     ... OK
## * checking whether the package can be unloaded cleanly ... OK
## * checking whether the namespace can be loaded with stated dependencies
##     ... OK
## * checking whether the namespace can be unloaded cleanly ... OK
## * checking loading without being on the library search path ... OK
## * checking dependencies in R code ... OK
## * checking S3 generic/method consistency ... OK
## * checking replacement functions ... OK
## * checking foreign function calls ... OK
## * checking R code for possible problems ... NOTE
## ggplot.lm: no visible global function definition for 'formula'
## ggplot.lm: no visible global function definition for 'predict'
## ggplot.lm: no visible binding for global variable 'fit'
## ggplot.lm: no visible global function definition for 'qnorm'
## ggplot.lm: no visible binding for global variable 'se.fit'
## meanPlot: no visible global function definition for 'plot'
## meanPlot: no visible global function definition for 'points'
## show,textplot: no visible global function definition for 'head'
## Undefined global functions or variables:
##    fit formula head plot points predict qnorm se.fit
```

```
## Consider adding
##    importFrom("graphics", "plot", "points")
##    importFrom("stats", "formula", "predict", "qnorm")
##    importFrom("utils", "head")
## to your NAMESPACE file.
## * checking Rd files ... OK
## * checking Rd metadata ... OK
## * checking Rd line widths ... OK
## * checking Rd cross-references ... OK
## * checking for missing documentation entries ... OK
## * checking for code/documentation mismatches ... OK
## * checking Rd \usage sections ... OK
## * checking Rd contents ... OK
## * checking for unstated dependencies in examples ... OK
## * checking contents of 'data' directory ... OK
## * checking data for non-ASCII characters ... OK
## * checking data for ASCII and uncompressed saves ... OK
## * checking examples ... ERROR
## Running examples in 'AdvancedRPkg-Ex.R' failed
## The error most likely occurred in:
##
## > base::assign(".ptime", proc.time(), pos = "CheckExEnv")
## > ### Name: ggplot.lm
## > ### Title: Method for plotting linear models
## > ### Aliases: ggplot.lm
## >
## > ### ** Examples
## >
## > ggplot(
## +    lm(mpg ~ hp * qsec, data = mtcars),
## +    aes(hp, mpg, linetype = factor(qsec)),
## +    vars = list(
## +      hp = min(mtcars$hp):max(mtcars$hp),
## +      qsec = round(mean(mtcars$qsec) + c(-1, 1) * sd(mtcars$qsec)), 1)) +
## +    geom_ribbon(aes(ymin = LL, ymax = UL), alpha = .2) +
```

```
## +    geom_line() +
## +    theme_bw()
## Error in ggplot(lm(mpg ~ hp * qsec, data = mtcars), aes(hp, mpg,
    linetype = factor(qsec)),
:
##    could not find function "ggplot"
## Execution halted
## * checking for unstated dependencies in 'tests' ... OK
## * checking tests ...
## Running 'testthat.R'
## OK
## * checking for detritus in the temp directory ... OK
## * DONE
##
## Status: 1 ERROR, 1 WARNING, 1 NOTE
## See
##    'C:/.../Temp/.../AdvancedRPkg.Rcheck/00check.log'
## for details.
##
## -- R CMD check results ----------------- AdvancedRPkg 0.0.0.9000 ----
## Duration: 28s
##
## > checking examples ... ERROR
##    Running examples in 'AdvancedRPkg-Ex.R' failed
##    The error most likely occurred in:
##
##    > base::assign(".ptime", proc.time(), pos = "CheckExEnv")
##    > ### Name: ggplot.lm
##    > ### Title: Method for plotting linear models
##    > ### Aliases: ggplot.lm
##    >
##    > ### ** Examples
##    >
##    > ggplot(
## +    lm(mpg ~ hp * qsec, data = mtcars),
## +    aes(hp, mpg, linetype = factor(qsec)),
```

```
##   +   vars = list(
##   +       hp = min(mtcars$hp):max(mtcars$hp),
##   +       qsec = round(mean(mtcars$qsec) + c(-1, 1) * sd(mtcars$qsec)),
            1)) +
##   +   geom_ribbon(aes(ymin = LL, ymax = UL), alpha = .2) +
##   +   geom_line() +
##   +   theme_bw()
##   Error in ggplot(lm(mpg ~ hp * qsec, data = mtcars), aes(hp, mpg,
     linetype = factor(qsec)),
 :
##     could not find function "ggplot"
##   Execution halted
##
## > checking DESCRIPTION meta-information ... WARNING
##   Dependence on R version '3.6.3' not with patchlevel 0
##
## > checking R code for possible problems ... NOTE
##   ggplot.lm: no visible global function definition for 'formula'
##   ggplot.lm: no visible global function definition for 'predict'
##   ggplot.lm: no visible binding for global variable 'fit'
##   ggplot.lm: no visible global function definition for 'qnorm'
##   ggplot.lm: no visible binding for global variable 'se.fit'
##   meanPlot: no visible global function definition for 'plot'
##   meanPlot: no visible global function definition for 'points'
##   show,textplot: no visible global function definition for 'head'
##   Undefined global functions or variables:
##     fit formula head plot points predict qnorm se.fit
##   Consider adding
##     importFrom("graphics", "plot", "points")
##     importFrom("stats", "formula", "predict", "qnorm")
##     importFrom("utils", "head")
##   to your NAMESPACE file.
##
## 1 error x | 1 warning x | 1 note x
```

The check() function includes several steps. It first ensures the documentation is up-to-date, then builds a source package tar ball, and then runs Rcmd check on it. We can see here quite a few issues are caught by the checks. All of them essentially boil down to not telling R exactly what functions or packages are imported or required. Most issues arise because of differences between how packages and interactive R work. During an interactive session, the base, methods, graphics, utils, and stats packages are loaded by default, even without calling library(stats), for example. This means that by default in an interactive session, many functions are available on the search path. In packages, R has become stricter in recent versions and requires required functions outside the base package to be explicitly imported. The checks even suggest what we could add to our namespace. We use the roxygen2 package, so rather than adding those to our namespace, we add them to the R code files. Specifically, to AdvancedRPkg/R/plot_functions.R, right after the @export roxygen2 statements, we add this for meanPlot():

```
##' @importFrom graphics plot points
##' @importFrom stats formula
```

And this for the ggplot.lm() method in the same file:

```
##' @importFrom stats predict qnorm
```

To AdvancedRPkg/R/textplot.R, right after the @param object roxygen2 statement, we add this for the show() method:

```
##' @importFrom utils head
```

In addition, a similar problem occurs with no visible bindings for some global variables. This comes from the ggplot.lm() method; because it uses within(), R cannot tell that the variables have been defined, even though they are looked up within a data frame environment that contains them. Although this is a somewhat spurious note, it is good to get rid of all notes. One solution is to add a call to globalVariables() into our R files somewhere. At the top of AdvancedRPkg/R/plot_functions.R, we add the following:

```
globalVariables(c("fit", "se.fit"))
```

The last issue noted is ggplot() could not be found in one of the examples. Because our package imports the ggplot2 package, the ggplot() function is available to code within our package. However, if a user loads only our package, ggplot2 functions are not loaded for the user's search path. Examples for functions are run as users, and so the function is not available, and R throws an error. The best path forward here is complex. We could omit the example, but then the documentation is less helpful. We could move ggplot2 from the Imports field of the DESCRIPTION file to the Depends field. Packages depended on are loaded before loading a package. The downside of this approach is it forces users of our package to have ggplot2 loaded. This increases the odds of function masking for users because it forces many packages to be loaded. We could also ensure the code is not run. Finally, we could explicitly point to the ggplot2 package, by adding ggplot2::function(). This is cumbersome, as we use several ggplot2 functions in that example, and it needs to be added throughout. However, this ensures that the example works and avoids loading ggplot2 onto users' search path. Users are, of course, free to load ggplot2 should they wish by explicitly calling library(ggplot2) themselves. In the AdvancedRPkg/R/plot_functions.R file, we edit the roxygen2 code for the example for ggplot.lm(), as shown here:

```
##' @examples
##' ggplot2::ggplot(
##'    lm(mpg ~ hp * qsec, data = mtcars),
##'    ggplot2::aes(hp, mpg, linetype = factor(qsec)),
##'    vars = list(
##'       hp = min(mtcars$hp):max(mtcars$hp),
##'       qsec = round(mean(mtcars$qsec) + c(-1, 1) * sd(mtcars$qsec)), 1)) +
##'    ggplot2::geom_ribbon(ggplot2::aes(ymin = LL, ymax = UL), alpha = .2)
+
##'    ggplot2::geom_line() +
##'    ggplot2::theme_bw()
```

Next, we rerun the check() function:

```
check(cran = TRUE)

## Updating AdvancedRPkg documentation
## Writing NAMESPACE
## Loading AdvancedRPkg
## Writing NAMESPACE
## -- Building ---------------------------------------- AdvancedRPkg --
## Setting env vars:
## * CFLAGS    : -Wall -pedantic
```

```
## * CXXFLAGS  : -Wall -pedantic
## * CXX11FLAGS: -Wall -pedantic
## -------------------------------------------------------------------------
## v   checking for file 'C:\...\RCode\AdvancedRPkg/DESCRIPTION'
## -   preparing 'AdvancedRPkg':
## v   checking DESCRIPTION meta-information
## -   checking for LF line-endings in source and make files and shell
##     scripts
## -   checking for empty or unneeded directories
## -   looking to see if a 'data/datalist' file should be added
## -   building 'AdvancedRPkg_0.0.0.9000.tar.gz'
##
## -- Checking ---------------------------------------- AdvancedRPkg --
## Setting env vars:
## * _R_CHECK_CRAN_INCOMING_USE_ASPELL_: TRUE
## * _R_CHECK_CRAN_INCOMING_REMOTE_    : FALSE
## * _R_CHECK_CRAN_INCOMING_           : FALSE
## * _R_CHECK_FORCE_SUGGESTS_          : FALSE
## * NOT_CRAN                          : true
## -- R CMD check ----------------------------------------------------------
## Warning in dir.create(pkgoutdir, mode = "0755") :
##    'C:\...\Temp\...\AdvancedRPkg.Rcheck' already exists
## * using log directory 'C:/.../Temp/.../AdvancedRPkg.Rcheck'
## * using R version 3.6.3 (2020-02-29)
## * using platform: x86_64-w64-mingw32 (64-bit)
## * using session charset: ISO8859-1
## Warning in dir.create(dir, mode = "0755") :
##    'C:\...\Temp\...\AdvancedRPkg.Rcheck\00_pkg_src' already exists
## * using options '--no-manual --as-cran'
## * checking for file 'AdvancedRPkg/DESCRIPTION' ... OK
## * this is package 'AdvancedRPkg' version '0.0.0.9000'
## * package encoding: UTF-8
## * checking package namespace information ... OK
## * checking package dependencies ... OK
## * checking if this is a source package ... OK
```

```
## * checking if there is a namespace ... OK
## * checking for executable files ... OK
## * checking for hidden files and directories ... OK
## * checking for portable file names ... OK
## * checking whether package 'AdvancedRPkg' can be installed ... OK
## * checking installed package size ... OK
## * checking package directory ... OK
## * checking for future file timestamps ... OK
## * checking DESCRIPTION meta-information ... WARNING
## Dependence on R version '3.6.3' not with patchlevel 0
## * checking top-level files ... OK
## * checking for left-over files ... OK
## * checking index information ... OK
## * checking package subdirectories ... OK
## * checking R files for non-ASCII characters ... OK
## * checking R files for syntax errors ... OK
## * checking whether the package can be loaded ... OK
## * checking whether the package can be loaded with stated dependencies
##     ... OK
## * checking whether the package can be unloaded cleanly ... OK
## * checking whether the namespace can be loaded with stated dependencies
##     ... OK
## * checking whether the namespace can be unloaded cleanly ... OK
## * checking loading without being on the library search path ... OK
## * checking dependencies in R code ... OK
## * checking S3 generic/method consistency ... OK
## * checking replacement functions ... OK
## * checking foreign function calls ... OK
## * checking R code for possible problems ... OK
## * checking Rd files ... OK
## * checking Rd metadata ... OK
## * checking Rd line widths ... OK
## * checking Rd cross-references ... OK
## * checking for missing documentation entries ... OK
## * checking for code/documentation mismatches ... OK
```

```
## * checking Rd \usage sections ... OK
## * checking Rd contents ... OK
## * checking for unstated dependencies in examples ... OK
## * checking contents of 'data' directory ... OK
## * checking data for non-ASCII characters ... OK
## * checking data for ASCII and uncompressed saves ... OK
## * checking examples ... OK
## * checking for unstated dependencies in 'tests' ... OK
## * checking tests ...
##    Running 'testthat.R'
##   OK
## * checking for detritus in the temp directory ... OK
## * DONE
##
## Status: 1 WARNING
## See
##   'C:/.../Temp/.../AdvancedRPkg.Rcheck/00check.log'
## for details.
##
##
## -- R CMD check results ----------------- AdvancedRPkg 0.0.0.9000 ----
## Duration: 26.3s
##
## > checking DESCRIPTION meta-information ... WARNING
##    Dependence on R version '3.6.3' not with patchlevel 0
##
## -- Test failures --------------------------------------- testthat ----
##
## > library(testthat)
## > library(AdvancedRPkg)
## Error in library(AdvancedRPkg) :
##    there is no package called 'AdvancedRPkg'
## Execution halted
##
## 0 errors v | 1 warning x | 0 notes v
```

We get a few warnings about temporary files already existing since we ran the code before. We also get an error when trying to run the tests because although we have created it, we have not yet installed AdvancedRPkg.

We can fix this by installing the package into our local library. Make sure you pick the location of the tar ball that was built:

```
build()
install.packages("../AdvancedRPkg_0.0.0.9000.tar.gz")
```

To resolve this issue around temporary files (if you have that), you can close and restart R. Note that you will need to reload the packages. Then recheck it. If all goes well, we should get a status of OK, as shown in the follow output. You should also get zero errors and warnings, and the tests should have run OK:

```
check(cran = TRUE)
```

```
## Updating AdvancedRPkg documentation
## Writing NAMESPACE
## Loading AdvancedRPkg
## Writing NAMESPACE
## -- Building -------------------------------------------- AdvancedRPkg --
## Setting env vars:
## * CFLAGS : -Wall -pedantic
## * CXXFLAGS : -Wall -pedantic
## * CXX11FLAGS: -Wall -pedantic
## ------------------------------------------------------------------------
## v  checking for file 'c:\...\RCode\AdvancedRPkg/DESCRIPTION'
## -  preparing 'AdvancedRPkg':
## v  checking DESCRIPTION meta-information
## -  checking for LF line-endings in source and make files and shell
      scripts
## -  checking for empty or unneeded directories
## -  looking to see if a 'data/datalist' file should be added
## -  building 'AdvancedRPkg_0.0.0.9000.tar.gz'
##
## -- Checking -------------------------------------------- AdvancedRPkg --
## Setting env vars:
```

```
## * _R_CHECK_CRAN_INCOMING_USE_ASPELL_: TRUE
## * _R_CHECK_CRAN_INCOMING_REMOTE_    : FALSE
## * _R_CHECK_CRAN_INCOMING_           : FALSE
## * _R_CHECK_FORCE_SUGGESTS_          : FALSE
## * NOT_CRAN                          : true
## -- R CMD check -------------------------------------------------------
## * using log directory 'C:/.../Temp/.../AdvancedRPkg.Rcheck'
## * using R version 3.6.3 (2020-02-29)
## * using platform: x86_64-w64-mingw32 (64-bit)
## * using session charset: ISO8859-1
## * using options '--no-manual --as-cran'
## * checking for file 'AdvancedRPkg/DESCRIPTION' ... OK
## * this is package 'AdvancedRPkg' version '0.0.0.9000'
## * package encoding: UTF-8
## * checking package namespace information ... OK
## * checking package dependencies ... OK
## * checking if this is a source package ... OK
## * checking if there is a namespace ... OK
## * checking for executable files ... OK
## * checking for hidden files and directories ... OK
## * checking for portable file names ... OK
## * checking whether package 'AdvancedRPkg' can be installed ... OK
## * checking installed package size ... OK
## * checking package directory ... OK
## * checking for future file timestamps ... OK
## * checking DESCRIPTION meta-information ... OK
## * checking top-level files ... OK
## * checking for left-over files ... OK
## * checking index information ... OK
## * checking package subdirectories ... OK
## * checking R files for non-ASCII characters ... OK
## * checking R files for syntax errors ... OK
## * checking whether the package can be loaded ... OK
## * checking whether the package can be loaded with stated dependencies
##    ... OK
```

```
## * checking whether the package can be unloaded cleanly ... OK
## * checking whether the namespace can be loaded with stated dependencies
##     ... OK
## * checking whether the namespace can be unloaded cleanly ... OK
## * checking dependencies in R code ... OK
## * checking S3 generic/method consistency ... OK
## * checking replacement functions ... OK
## * checking foreign function calls ... OK
## * checking R code for possible problems ... OK
## * checking Rd files ... OK
## * checking Rd metadata ... OK
## * checking Rd line widths ... OK
## * checking Rd cross-references ... OK
## * checking for missing documentation entries ... OK
## * checking for code/documentation mismatches ... OK
## * checking Rd \usage sections ... OK
## * checking Rd contents ... OK
## * checking for unstated dependencies in examples ... OK
## * checking contents of 'data' directory ... OK
## * checking data for non-ASCII characters ... OK
## * checking data for ASCII and uncompressed saves ... OK
## * checking examples ... OK
## * checking for unstated dependencies in 'tests' ... OK
## * checking tests ...
##   Running 'testthat.R'
##  OK
## * checking for detritus in the temp directory ... OK
## * DONE
##
## Status: OK
##
## -- R CMD check results -------------------- AdvancedRPkg 0.0.0.9000 ----
## Duration: 26.8s
##
## 0 errors v | 0 warnings v | 0 notes v
```

At this point, we can commit all our changes to the Git repository and sync it with GitHub.

Now, should you wish, `library(AdvancedRPkg)` works in R.

You can also share the compressed tar ball created from the devtools package from `build()`. If you put the package on GitHub, it can alternately be installed readily by running the following code, replacing JWiley with your username and running this code at the R console (not the OS terminal!):

```
library(devtools)
install_github("JWiley/AdvancedRPkg")
```

If you want, you can edit the README file for the package. It is not required, but can be a useful reference. It is written using the Markdown markup language. If edited, it also shows up on GitHub. We edit the README.md file and add the text and markup that follows. Briefly, Markdown uses various numbers of hashes (#) to indicate header levels: # for level 1 and ## for level 2, for example. Text between asterisks, *, is emphasized (italicized). Brackets, [ ], and parentheses, ( ), are used to add URL links, and triple back ticks (") are used to show the start and end of a block of code:

```
# AdvancedRPkg
```

```
This is a sample \texttt{R} package that acompanies Chapter 6 of
*Advanced R 4.0.0: Data Programming and the Cloud*. To learn more,
check out the [book](\url{https://www.apress.com/us/book/9781484259726})
```

```
## Installation
```

```
You can install and test the package by running:
```

```
library(devtools)
install_github("JWiley/AdvancedRPkg")
```

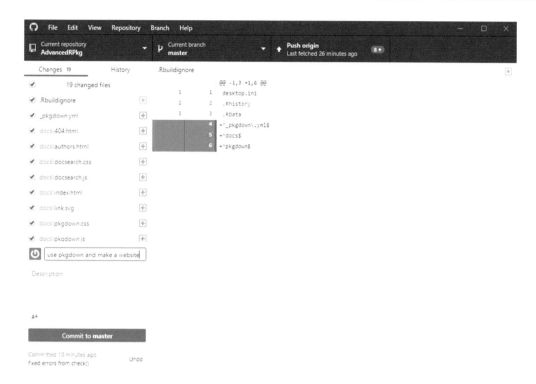

Figure 6-12. *Commit changes from building a package website*

6.5 R Package Website

We can use the pkgdown package to build a nice website for our package that will sit
under the docs folder. First, to configure our package to use pkgdown, we again rely on
the usethis package:

```
use_pkgdown(config_file = "_pkgdown.yml", destdir = "docs")
## v Setting active project to 'C:/.../RCode/AdvancedRPkg'
## v Adding '^_pkgdown\\.yml$', '^docs$' to '.Rbuildignore'
## v Adding '^pkgdown$' to '.Rbuildignore'
## v Writing '_pkgdown.yml'
## * Modify '_pkgdown.yml'
```

You can modify the file `pkgdown.yml`, but it is not required. If you want to customize things, see `https://pkgdown.r-lib.org/` for more details on what you can do with the `pkgdown` package. We will use the defaults for now. We can now build the package website by running `build site()`:

```
build_site()
```

This should not produce any output if it all works. You will get the main page popup in your browser. To get the package hosted on GitHub along with our website for the package, we need to commit all the changes and push. There are a lot of files created for the package website as shown in Figure 6-12.

Now, we can see all our changes online: `http://github.com/JWiley/AdvancedRPkg`. If GitHub is properly configured, you can also view the website online. To configure GitHub, go to the settings for the repository and choose GitHub Pages with the sources as the `/docs` folder as shown in Figure 6-13.

Because we have a custom URL set up and linked to our GitHub account, you can access the package website at `http://joshuawiley.com/AdvancedRPkg`.

However, you can have a website even without a custom domain. It will be `http://jwiley.github.io/AdvancedRPkg`, and you just need to replace `jwiley` with your GitHub username.

Finally, you can submit your package to CRAN should you wish. The first step is to carefully attend to all of their current policies, located at the *CRAN Repository* Policy site (`https://cran.r-project.org/web/packages/policies.html`). Once you have checked that your package complies and have corrected any noncompliance, you can submit it to CRAN at `https://cran.r-project.org/submit.html`. CRAN is huge, with thousands of packages and a tremendous number of updates daily. It is a free service to the community run by volunteers. Thus, even if some of the requirements are tedious, if you want your package on CRAN, it is only fair to play by their rules and do whatever makes it easiest for them. Otherwise, GitHub is a relatively easy place to host and distribute package source code.

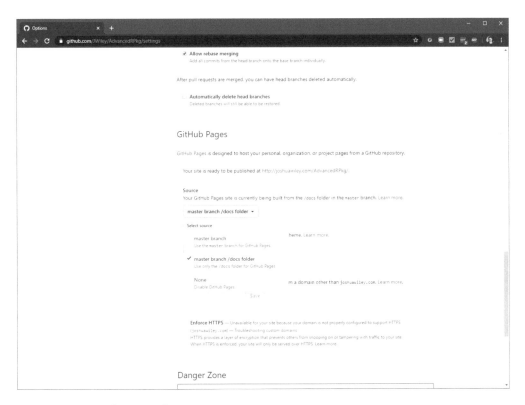

Figure 6-13. *Package website*

6.6 Summary

This chapter covered the logistics of developing, testing, documenting, and releasing an R package. Although the process does not necessarily involve complex R code, handling the many aspects and getting all pieces to interact properly can be challenging. The payoff for the work is ease of installation and use for users, along with high-quality documentation and assurances that the code works as intended. If you plan to continue developing packages, useful resources for further reading are the official manual, "Writing R Extensions" (`https://cran.r-project.org/doc/manuals/R-exts.html`), and documentation for using `roxygen2` (`https://cran.r-project.org/web/packages/roxygen2/vignettes/roxygen2.html`). The vignettes for `roxygen2` are especially useful for topics not covered in this chapter, such as formatting the documentation and collation order (required when some classes or functions have to be loaded before others in your package). A brief summary of the functions used in this chapter is shown in Table 6-3.

This chapter is also the last chapter focused specifically on R programming and the tools around software development in R. The remainder of this book focuses on using R for data management and applied analysis at an advanced level. Although we utilize many aspects of the R programming you learned in this section of the book and write many functions, we do not develop any new packages in the upcoming chapters.

Table 6-3. *Listing of key functions described in this chapter and summary of what they do*

Function	What It Does
setup()	Creates a new package (in our case, AdvancedRPkg).
list.files()	Does what it says—the R equivalent of the Unix or Windows ls.
load_all()	Before a package is built or installed, library() does not work; this simulates that in the meantime.
context()	Part of the testthat package we used in test_textplot.R.
test_that()	Part of the testthat package we used in test_textplot.R.
expect_*()	A family of functions useful for testing code. These functions check that output matches a certain expectation, such as the class of output, whether the code returns an error or warning, and many others. For a full list, after loading the testthat package, run apropos("expect_") at the console.
apropos()	Useful to search for partially remembered function calls.
test()	Runs the testthat tests; run from our chapter06.R file.
package_coverage()	Calculates how much of the package has been tested; usually, the goal is 100 percent.
document()	Used to build the roxygen2 documentation for an R package.
build()	Used to build the R package.
check()	Used to check the R package; we used the CRAN option in ours.
build_site()	Used to use pkgdown to build a website for our package.

CHAPTER 7

Introduction to `data.table`

We already briefly introduced the data.table package. This package is the heart of this chapter, which covers the basics of accessing, editing, and manipulating data under the broad term data management. Although not glamorous, data management is a critical first step to data visualization or analysis. Furthermore, the majority of time on a particular analysis project may come from the data management. For example, running a linear model in R can take one line of code, once the data is clean and in the format that the `lm()` function in R expects. Data management can be challenging, because raw data come in all types, shapes, and formats; missing data is common; and you may also have to combine or merge separate data sources. In this chapter, we use the data.table [9] and extraoperators [35] R packages:

```
options(width = 70)
library(data.table)
library(extraoperators)
```

7.1 Introduction to `data.table`

One of the benefits of using data tables, rather than the built-in data frame class objects, is data tables are more memory efficient and faster to manipulate and modify. The `data.table` package accomplishes this to a large extent by altering data tables in place in memory, whereas with data frames, R typically makes a copy of the data, modifies it, and stores it. Making a full copy happens regardless of whether all columns of the data are being changed or, for example, 1 of 100 columns is being changed. Consequently, operations on data frames tend to take more time and use up more memory than

© Matt Wiley and Joshua F. Wiley 2020
M. Wiley and J. F. Wiley, *Advanced R 4 Data Programming and the Cloud*,
https://doi.org/10.1007/978-1-4842-5973-3_7

comparable operations with data tables. To show the concepts of data management, we work with small data, so memory and processor time are not an issue. However, we highlight the use of data tables, rather than data frames, because data tables scale gracefully to far larger amounts of data than do data frames. There are other benefits of data tables as well, including not requiring all variable names to be quoted, that we show throughout the chapter. In contrast, the major advantage of data frames is that they live in base R and more people are familiar with working with them (which is helpful, e.g., if you share your code).

Data tables can be quite large. Rather than viewing all your data, it is often helpful to view just the head() and the tail(). When you show a data table in R by typing its name, the first and last five rows are returned by default. R also defaults to printing seven significant digits, which can make our numerical entries messy. For cleanness, we use the options() function to control some of the global options so that only the first and last three rows print (if the data table has more than 20 rows total—otherwise, all are shown); in addition, only two significant digits display for our numerical entries. While we set this, we also introduce Edgar Anderson's iris data. This data involves sepal and petal lengths and widths in centimeters of three species of iris flowers.

Before we convert the iris data into a data table, we are going to convert our species' names to a character rather than factor class. In the following code, we set our options, force the species' names to characters, and create our data table, diris:

```
options(stringsAsFactors = FALSE,
        datatable.print.nrows = 20, #if over 20 rows
        datatable.print.topn = 3, #print first and last 3
        digits = 2) #reduce digits printed by R

iris$Species <- as.character(iris$Species)
diris <- as.data.table(iris)

diris
```

```
##      Sepal.Length Sepal.Width Petal.Length Petal.Width   Species
## 1:            5.1         3.5          1.4         0.2    setosa
## 2:            4.9         3.0          1.4         0.2    setosa
## 3:            4.7         3.2          1.3         0.2    setosa
## ---
```

## 148:	6.5	3.0	5.2	2.0 virginica
## 149:	6.2	3.4	5.4	2.3 virginica
## 150:	5.9	3.0	5.1	1.8 virginica

There are 150 rows of data comprising our three species: setosa, versicolor, and virginica. Each species has 50 measurement sets. Keeping in mind R lives in memory, it is important to note huge data tables may cause issues. To check for space, the tables() command shows all the data tables in memory, as well as their sizes. In fact, tables() is itself a data table. Notice the function gives information as to the name of our data table(s), the number of rows and columns, the size in MB, and if there is a key. We discuss data tables with a key in just a few paragraphs. For now, 1MB is not likely to give us memory issues:

```
tables()
```

```
##      NAME NROW NCOL MB
## 1: diris  150    5  0
##                                              COLS KEY
## 1: Sepal.Length,Sepal.Width,Petal.Length,Petal.Width,Species
## Total: 0MB
```

If we need more information about the types of data in each column, calling the sapply() command on a data table of interest along with the class() function works. Recall that the *apply functions return results from calling the function named in the second formal on the elements from the first formal argument. In our case, class() is called on diris, and we see what types of values are stored in the columns of diris:

```
sapply(diris, class)
```

```
## Sepal.Length  Sepal.Width  Petal.Length  Petal.Width       Species
##    "numeric"    "numeric"     "numeric"     "numeric"   "character"
```

Data tables have a powerful conceptual advantage over other data formats. Namely, data tables may be keyed. With a key, it is possible to use binary search. For those new to binary search, imagine needing to find a name in a phone book. It would take quite some time to read every name. However, because we know the telephone directory sorts alphabetically, we can instantly find a middle-of-the-alphabet name, determine whether our search name belongs before or after that name, and remove half the phone book from our search in the process. Our algorithm may be repeated with the remaining half, leaving only a quarter of the original data to search after just two accesses! More mathematically, while a search through all terms of a data frame would be linear based on the number of rows, n, in a data table, it is at worse log n. To make this tangible, for some tables we use, doubling our data added only minutes to code runtime rather than doubling runtime—a major win.

Note   In data.table, a key is an index that may or may not be unique, created from one or more columns in the data. The data is sorted by the key, allowing very fast operations using the key. Example operations include subsetting or filtering, performing a calculation (e.g., the mean of a variable) for every unique key value, and merging two or more data sets.

Of course, it does take time to key a data table, so it depends on how often you access your data before making this sort of choice. Another consideration is what to use as the key. A table has only one key, although that key may consist of more than one column. The function setkey() takes a data table as the first formal and then takes an unspecified number of column names after that to set a key. To see whether a data table has a key, simply call haskey() on the data table you wish to test. Finally, if you want to know what columns built the key, the function key() called on the data table provides that information. We show these commands on our data table diris in the following code lines:

```
haskey(diris)
```

```
## [1] FALSE
```

```
setkey(diris, Species)
key(diris)
```

```
## [1] "Species"
```

```
haskey(diris)
```

```
## [1] TRUE
```

Keys often are created based on identification variables. For instance, in research, each participant in a study may be assigned a unique ID. In business cases, every customer may have an ID number. Data tables often are keyed by IDs like these, because often operations are performed by those IDs. For example, when conducting research in a medical setting, you may have two data tables, one containing questionnaires completed by participants and another containing data from medical records. The two tables can be joined by participant ID. In business, every time a customer makes a new purchase, an additional row may be added to a data set containing the purchase amount and customer ID. In addition to knowing individual purchases, however, it may also be helpful to know how much each customer has purchased in total—the sum purchase amount for each customer ID.

It is not required to set a key; the essence of binary sort simply requires a logical order. The function call order() does this. It is important to note that data tables have another nice feature. Going back to our phone book analogy, imagine that we want to sort by first name instead of last name. Now, the complicated way to do this would be to line up every person in the phone book and then have them move into the new order. In this example, the people are the data. However, it would be much easier simply to reorganize the phone book—much less data. The telephone directory is a reference to the physical location of the people. We can simply move some of those entries higher up in our reference table, without ever changing anyone's physical address. It is again a faster technique that avoids deleting and resaving data in memory. So too, when we impose order on a data table, it is by reference. Note this new order unsets our key, since a critical aspect of a key is that the data is sorted by that key:

```
diris <- diris[order(Sepal.Length)]
diris
```

```
##      Sepal.Length Sepal.Width Petal.Length Petal.Width   Species
##   1:          4.3         3.0          1.1         0.1    setosa
##   2:          4.4         2.9          1.4         0.2    setosa
##   3:          4.4         3.0          1.3         0.2    setosa
## ---
## 148:          7.7         2.8          6.7         2.0 virginica
## 149:          7.7         3.0          6.1         2.3 virginica
## 150:          7.9         3.8          6.4         2.0 virginica
```

```
haskey(diris)
```

```
## [1] FALSE
```

Alternatively, we may order a data table by using multiple variables and even change from the default increasing order to decreasing order. To use more than one column, simply call order() on more than column. Adding - before a column name sorts in decreasing order. The following code sorts diris based on increasing sepal length and then decreasing sepal width:

```
diris <- diris[order(Sepal.Length, -Sepal.Width)]
diris
```

```
##         Sepal.Length Sepal.Width Petal.Length Petal.Width   Species
##   1:             4.3         3.0          1.1         0.1    setosa
##   2:             4.4         3.2          1.3         0.2    setosa
##   3:             4.4         3.0          1.3         0.2    setosa
## ---
## 148:             7.7         2.8          6.7         2.0 virginica
## 149:             7.7         2.6          6.9         2.3 virginica
## 150:             7.9         3.8          6.4         2.0 virginica
```

If we reset our key, it destroys our order. While this would not be apparent with our normal call to diris, if we take a look at just the correct rows, we lose the strict increasing order on sepal length:

```
setkey(diris, Species)
diris[49:52]
```

```
##     Sepal.Length Sepal.Width Petal.Length Petal.Width   Species
## 1:           5.7         3.8          1.7         0.3    setosa
## 2:           5.8         4.0          1.2         0.2    setosa
## 3:           4.9         2.4          3.3         1.0 versicolor
## 4:           5.0         2.3          3.3         1.0 versicolor
```

This teaches us that decisions about order and keys come with a cost. Part of the improved speed of a data table comes with being able to perform fast searches. Choosing a key comes with a one-time computation cost, so it behooves us to make sensible choices for our key. The iris data set is small enough to make such costs irrelevant while we learn, but real-life data sets are not as likely to be so forgiving. Choosing a key also influences some of the default choices for functions we use to understand aspects of our data table. Setting a key is about not only providing a sort of our data but also asserting our intent to make the key the determining aspect of that data. An example of this is shown with the anyDuplicated() function call, which gives the index of the first duplicated row. As we have set our key to Species, it defaults to look in just that column. Notice the first duplicate for Species occurs in the second row, while the first duplicated sepal length entry takes place in the third row:

```
anyDuplicated(diris)
```

```
## [1] 106
```

```
anyDuplicated(diris$Sepal.Length)
```

```
## [1] 3
```

Other functions in the duplication family of functions that can be influenced by key include duplicated() itself, which returns a boolean value for each row. Now by default in more modern versions of data.table, duplicated() checks all elements of a row against prior rows. Please note this is a change for how duplicated() worked in the past.

That said, through use of a second formal, by = key(), this function will respect key values. In this case, such a call on diris would give us FALSE values for each of the rows 1, 51, and 101, since we have 50 of each species and are sorting by species, so we instead put that into a nice table:

```
haskey(diris)
```

```
## [1] TRUE
```

```
key(diris)
```

```
## [1] "Species"
```

```
table(duplicated(diris)) # not respecting key
```

```
##
## FALSE   TRUE
##   149      1
```

```
table(duplicated(diris, by = key(diris))) # respecting key
```

```
##
## FALSE   TRUE
##     3    147
```

Similarly, the modern behavior of the `unique()` function defaults to ignoring key and would only remove row 79 from our data set. However, by using `by = key()`, only the first instance of each new key would appear:

```
unique(diris, by = key(diris))
```

```
##     Sepal.Length Sepal.Width Petal.Length Petal.Width    Species
## 1:           4.3         3.0          1.1         0.1     setosa
## 2:           4.9         2.4          3.3         1.0 versicolor
## 3:           4.9         2.5          4.5         1.7  virginica
```

One of the benefits of having a data table is the ability to define a key. The key should be chosen to represent the true uniqueness of a data table. Now, in some types of data, the key might be obvious. Analysis of employee records would likely use employee identity numbers or perhaps government taxpayer identity numbers. However, in other scenarios such as this one, the key might not be so obvious. Indeed, as we see in the preceding results of the `unique()` call on our keyed `diris` data table, we get only three rows returned. That may not truly represent any genuine uniqueness. In that case, it may benefit us to have more than one column used to create the key, although which columns are relevant depends entirely on the data available and the goals of our final analysis.

7.2 Selecting and Subsetting Data
Using the First Formal

Data tables were built to easily and quickly access data. Part of the way we achieve this is by not having formal row names. Indeed, part of what a key does could be considered providing a useful row name rather than an arbitrary index. The overall format for data table objects is of the form `Data.Table[i, j, by]`, where `i` expects row information. In particular, data tables are somewhat self-aware in that they understand their column names as variables without using the `$` operator. If a column variable is not named, then it is imagined we are attempting to call by row:

Note   Rows from a data table are commonly selected by passing a vector containing one of the following: (1) numbers indicating the rows to choose, d[1:5]; (2) negative numbers indicating the rows to exclude, d[-(1:5)]; and (3) a logical vector indicating whether each row should be included, d[v == 1], often as a logical test using a column, v, from the data set.

```
diris[1:5]
```

```
##    Sepal.Length Sepal.Width Petal.Length Petal.Width Species
## 1:          4.3         3.0          1.1         0.1  setosa
## 2:          4.4         3.2          1.3         0.2  setosa
## 3:          4.4         3.0          1.3         0.2  setosa
## 4:          4.4         2.9          1.4         0.2  setosa
## 5:          4.5         2.3          1.3         0.3  setosa
```

Notice that if this were a data frame, we would have gotten the entire data frame, because we would have been calling for columns and we have only five columns. Here, in data table world, because we have not called for a specific column name, we get the first five rows instead. As with increasing or decreasing, we can also perform an opposite selection via the - operator. In this case, by asserting we want the complement of rows 1:148, we get just the last two rows. Notice the row numbers on the left side. These are rows 149 and 150, yet, because the data table is not concerned with row names, they are renumbered on the fly as they print to our screen. It is wise to use caution when selecting rows in a data table:

```
diris[-(1:148)]
```

```
##    Sepal.Length Sepal.Width Petal.Length Petal.Width   Species
## 1:          7.7         2.6          6.9         2.3 virginica
## 2:          7.9         3.8          6.4         2.0 virginica
```

More typically, we select by column name or by key. As we have set our key to Species, it becomes very easy to select just one species. We simply ask for the character string to match. Now, at the start of this chapter, we made sure species were character strings. The reason for that choice is clear in the following code, where being able to readily select our desired key values is quite handy. Notice the second command uses the negation operator, !, to assert we want to select all key elements that are not part of the character string we typed:

```
diris["setosa"]
```

```
##       Sepal.Length Sepal.Width Petal.Length Petal.Width Species
##   1:           4.3         3.0          1.1         0.1  setosa
##   2:           4.4         3.2          1.3         0.2  setosa
##   3:           4.4         3.0          1.3         0.2  setosa
## ---
## 48:           5.7         4.4          1.5         0.4  setosa
## 49:           5.7         3.8          1.7         0.3  setosa
## 50:           5.8         4.0          1.2         0.2  setosa
```

```
diris[!"setosa"]
```

```
##       Sepal.Length Sepal.Width Petal.Length Petal.Width    Species
##    1:          4.9         2.4          3.3         1.0 versicolor
##    2:          5.0         2.3          3.3         1.0 versicolor
##    3:          5.0         2.0          3.5         1.0 versicolor
## ---
##  98:          7.7         2.8          6.7         2.0  virginica
##  99:          7.7         2.6          6.9         2.3  virginica
## 100:          7.9         3.8          6.4         2.0  virginica
```

Of course, while we have chosen so far to use key elements, that is by no means required. Logical indexing is quite natural with any of our columns. The overall format compares elements of the named column to the logical tests set up. The final result shows only the subset of rows that evaluated to TRUE. For inequalities, multiple arguments, and strict equality (in essence, the usual comparison operators), data tables work as you would expect. Also, note again with data tables, we do not need to reference in which object the variable Sepal.Length is. It evaluates within the data table by default:

```
diris[Sepal.Length < 5]
```

```
##       Sepal.Length Sepal.Width Petal.Length Petal.Width   Species
##   1:           4.3         3.0          1.1         0.1    setosa
##   2:           4.4         3.2          1.3         0.2    setosa
##   3:           4.4         3.0          1.3         0.2    setosa
## ---
```

## 20:	4.9	3.0	1.4	0.2	setosa
## 21:	4.9	2.4	3.3	1.0	versicolor
## 22:	4.9	2.5	4.5	1.7	virginica

```
diris[Sepal.Length < 5 & Petal.Width < .2]
```

##	Sepal.Length	Sepal.Width	Petal.Length	Petal.Width	Species
## 1:	4.3	3.0	1.1	0.1	setosa
## 2:	4.8	3.0	1.4	0.1	setosa
## 3:	4.9	3.6	1.4	0.1	setosa
## 4:	4.9	3.1	1.5	0.1	setosa

```
diris[Sepal.Length == 4.3 | Sepal.Length == 4.4]
```

##	Sepal.Length	Sepal.Width	Petal.Length	Petal.Width	Species
## 1:	4.3	3.0	1.1	0.1	setosa
## 2:	4.4	3.2	1.3	0.2	setosa
## 3:	4.4	3.0	1.3	0.2	setosa
## 4:	4.4	2.9	1.4	0.2	setosa

There are, however, some new operators. In the preceding code, we typed a bit too much because we repeated Sepal.Length. Thinking forward to what data tables store, imagine we occasionally modify a list of employees. We might simply store an employee's unique identifier to that list and pull from our data table as needed. Of course, in our example here, we have sepal lengths rather than employees, and we want just the rows that have the lengths of interest in our list. The command is %in% and takes a column name on the left and a vector on the right. Naturally, we may want the complement of such a group, in which case we use the "not in" operator, %nin%. We show both results in the following code:

```
interest <- c(4.3, 4.4)
```

```
diris[Sepal.Length %in% interest]
```

##	Sepal.Length	Sepal.Width	Petal.Length	Petal.Width	Species
## 1:	4.3	3.0	1.1	0.1	setosa
## 2:	4.4	3.2	1.3	0.2	setosa
## 3:	4.4	3.0	1.3	0.2	setosa
## 4:	4.4	2.9	1.4	0.2	setosa

```
diris[Sepal.Length %nin% interest]
```

```
##      Sepal.Length Sepal.Width Petal.Length Petal.Width   Species
##   1:          4.5         2.3          1.3         0.3    setosa
##   2:          4.6         3.6          1.0         0.2    setosa
##   3:          4.6         3.4          1.4         0.3    setosa
## ---
## 144:          7.7         2.8          6.7         2.0 virginica
## 145:          7.7         2.6          6.9         2.3 virginica
## 146:          7.9         3.8          6.4         2.0 virginica
```

Using the Second Formal

Thus far, we have mentioned data tables do not have native row names, and yet, we have essentially been selecting specific rows in a variety of ways. Recall our generic layout Data.Table[i, j, by] allows us to select objects of columns in the first formal. We turn our attention to the second formal location of j, with a goal to select only some columns rather than all columns. Keep in mind, data tables presuppose variables passed within them exist in the environment of the data table. Thus, it becomes quite easy to ask for all the variables in a single column:

Note   Variables are selected by typing the unquoted variable name, d[, variable], or multiple variables separated by commas within, .(), d[, .(v1, v2)].

```
diris[, Sepal.Length]
```

```
##   [1] 4.3 4.4 4.4 4.4 4.5 4.6 4.6 4.6 4.6 4.7 4.7 4.8 4.8 4.8 4.8 4.8
##  [17] 4.9 4.9 4.9 4.9 5.0 5.0 5.0 5.0 5.0 5.0 5.0 5.0 5.1 5.1 5.1 5.1
##  [33] 5.1 5.1 5.1 5.1 5.2 5.2 5.2 5.3 5.4 5.4 5.4 5.4 5.4 5.5 5.5 5.7
##  [49] 5.7 5.8 4.9 5.0 5.0 5.1 5.2 5.4 5.5 5.5 5.5 5.5 5.5 5.6 5.6 5.6
##  [65] 5.6 5.6 5.7 5.7 5.7 5.7 5.7 5.8 5.8 5.8 5.9 5.9 6.0 6.0 6.0 6.0
##  [81] 6.1 6.1 6.1 6.1 6.2 6.2 6.3 6.3 6.3 6.4 6.4 6.5 6.6 6.6 6.7 6.7
##  [97] 6.7 6.8 6.9 7.0 4.9 5.6 5.7 5.8 5.8 5.8 5.9 6.0 6.0 6.1 6.1 6.2
## [113] 6.2 6.3 6.3 6.3 6.3 6.3 6.3 6.4 6.4 6.4 6.4 6.4 6.5 6.5 6.5 6.5
```

```
## [129] 6.7 6.7 6.7 6.7 6.7 6.8 6.8 6.9 6.9 6.9 7.1 7.2 7.2 7.2 7.3 7.4
## [145] 7.6 7.7 7.7 7.7 7.7 7.9
```

```
is.data.table(diris[,Sepal.Length])
```

```
## [1] FALSE
```

On the other hand, as you can see in the preceding vector return, entering a single column name does not return a data table. If a data table return is required, this may be solved in more than one way. Recall the data table strives to be both computationally fast and programmatically fast. We extract single or multiple columns while retaining the data table structure by accessing them in a list. This common desire to pass a list() as an argument for data has a clear, shorthand function call in the data table of .(). We show in the following code we may pass a list with one or more elements calling out specific column names. Both return data tables:

```
diris[, .(Sepal.Length, Sepal.Width)]
```

```
##       Sepal.Length Sepal.Width
##   1:          4.3         3.0
##   2:          4.4         3.2
##   3:          4.4         3.0
## ---
## 148:          7.7         2.8
## 149:          7.7         2.6
## 150:          7.9         3.8
```

```
diris[, .(Sepal.Length)]
```

```
##       Sepal.Length
##   1:          4.3
##   2:          4.4
##   3:          4.4
## ---
## 148:          7.7
## 149:          7.7
## 150:          7.9
```

Using the Second and Third Formals

We have been using the column names as variables that may be directly accessed. They evaluate within the frame of our data table diris. This feature may be turned on or off by the with argument that is one of our options for the last formal in our generic layout Data.Table[i, j, by]. The default value is with = TRUE, where default behavior is to evaluate the j formal within the frame of our data table diris. Changing the value to with = FALSE changes the default behavior. This difference is similar to what happens when working with R interactively. We assign the text, example, to a variable, x. If we type x without quotes at the R console, R looks for an object named x. If we type "x" in quotes, R treats it not as the name of an R object but as a character string:

```
x <- "example"
x

## [1] "example"

"x"

## [1] "x"
```

When with = TRUE (the default), data tables behave like the interactive R console. Unquoted variable names are searched for as variables within the data table. Quoted variable names or numbers are evaluated literally as vectors in their own right. When with = FALSE, data tables behave more like data frames, and instead of treating numbers or characters as vectors in their own right, they search for the column name or position that matches them. In this context then, our data table expects j to be either a column position in numeric form or a character vector of the column name(s). In both cases, data tables are still returned:

```
diris[, 1, with = FALSE]

##         Sepal.Length
##    1:            4.3
##    2:            4.4
##    3:            4.4
## ---
## 148:            7.7
## 149:            7.7
```

```
## 150:              7.9
```

```
diris[, "Sepal.Length", with = FALSE]
```

```
##       Sepal.Length
##    1:          4.3
##    2:          4.4
##    3:          4.4
## ---
## 148:          7.7
## 149:          7.7
## 150:          7.9
```

This becomes a very useful feature if columns need to be selected dynamically. Character strings pass to the data table via variables, such as in the following code:

```
v <- "Sepal.Length"
diris[, v, with = FALSE]
```

```
##       Sepal.Length
##    1:          4.3
##    2:          4.4
##    3:          4.4
## ---
## 148:          7.7
## 149:          7.7
## 150:          7.9
```

Of course, if we do not turn off the default behavior via the third formal, this may lead to unexpected results. Caution is required to ensure you get the data you seek. However, remember what each of these returns; the results that follow are not, in fact, unfortunate or mistakes. They prove to be quite useful:

```
diris[, v]
```

```
## Error in '[.data.table'(diris, , v): j (the 2nd argument inside [...])
is a single symbol but column
```

```
diris[, 1]
```

```
##       Sepal.Length
##    1:          4.3
##    2:          4.4
##    3:          4.4
## ---
## 148:          7.7
## 149:          7.7
## 150:          7.9
```

As you have seen, we may call out one or more columns in a variety of ways, depending on our needs. Recalling the useful notation of - to subtract from the data table, we are also able to remove one or more columns from our data table, giving us access to all the rest. Notice in particular that we could use our variable v to hold multiple columns if necessary, thus allowing us to readily keep a list of columns that need removing. For instance, this can be used to remove sensitive information in data that goes from being for internal use only to being available for external. In the following subtraction examples, we ask for the first row in all cases to save space and ink:

```
diris[1, -"Sepal.Length", with=FALSE]
```

```
##    Sepal.Width Petal.Length Petal.Width Species
## 1:           3          1.1         0.1 setosa
```

```
diris[1, !"Sepal.Length", with=FALSE]
```

```
##    Sepal.Width Petal.Length Petal.Width Species
## 1:           3          1.1         0.1 setosa
```

```
diris[1, -c("Sepal.Length", "Petal.Width"), with=FALSE]
```

```
##    Sepal.Width Petal.Length Species
## 1:           3          1.1 setosa
```

```
diris[1, -v, with=FALSE]
```

```
##    Sepal.Width Petal.Length Petal.Width Species
## 1:           3          1.1         0.1 setosa
```

Contrastingly, on occasion, we do want to return our column as a vector as with diris[,Sepal.Length]. In that case, there are easy-enough ways to gain access to that vector. Here, we combine all our function call examples with the head() function to truncate output. Notice that the second example is convenient if you need variable access to that column as a vector, while the last reminds us that data tables build on data frames:

```
head(diris[["Sepal.Length"]])
```

```
## [1] 4.3 4.4 4.4 4.4 4.5 4.6
```

```
head(diris[[v]])
```

```
## [1] 4.3 4.4 4.4 4.4 4.5 4.6
```

```
head(diris[, Sepal.Length])
```

```
## [1] 4.3 4.4 4.4 4.4 4.5 4.6
```

```
head(diris$Sepal.Length)
```

```
## [1] 4.3 4.4 4.4 4.4 4.5 4.6
```

7.3 Variable Renaming and Ordering

Data tables come with variables and a column order. On occasion, it may be desirable to change that order or rename variables. In this section, we show some techniques to do precisely that, keeping our comments brief and letting the code do the talking.

Note   Variable names in data tables can be viewed by using names(data), changed by using setnames(data, oldnames, newnames), and reordered by using setnames(data, neworder).

While we have not yet used this function in this book, names() is part of R's base code and is a generic function. Data tables have only column names, so the function call colnames() in the base package is also going to return the same results:

```
names(diris)
```

```
## [1] "Sepal.Length" "Sepal.Width" "Petal.Length" "Petal.Width"
## [5] "Species"
```

```
colnames(diris)
```

```
## [1] "Sepal.Length" "Sepal.Width" "Petal.Length" "Petal.Width"
## [5] "Species"
```

For objects in the base code of R, such as a data frame, names() can even be used to set new values. However, data tables work differently and use a different function to be memory efficient. Remember, data tables avoid copying data whenever possible and instead simply change the reference pointers. Normally, renaming an object in R would copy the data, which both takes time to accomplish and takes up space in memory. The function call for data tables is setnames(), which takes formal arguments of the data table, the old column name(s), and the new column name(s):

```
setnames(diris, old = "Sepal.Length", new = "SepalLength")
names(diris)
```

```
## [1] "SepalLength"  "Sepal.Width" "Petal.Length" "Petal.Width"
## [5] "Species"
```

We could have also referred to the old column by numeric position rather than variable name. This can be convenient when names are long or the data's position is known in advance:

```
setnames(diris, old = 1, new = "SepalL")
names(diris)
```

```
## [1] "SepalL"       "Sepal.Width" "Petal.Length" "Petal.Width"
## [5] "Species"
```

Additionally, columns may be fully reordered. Again, we prefix our command with set, as the command setcolorder() sets the column order while avoiding any copying of the underlying data. All that changes is the order of the pointers to each column. Thus, this is a fast command for data tables that depends on only the number of columns rather than on the number of elements within a column:

```
setcolorder(diris, c("SepalL", "Petal.Length", "Sepal.Width", "Petal.Width",
"Species"))
diris
```

```
##      SepalL Petal.Length Sepal.Width Petal.Width  Species
##   1:    4.3          1.1         3.0         0.1   setosa
##   2:    4.4          1.3         3.2         0.2   setosa
##   3:    4.4          1.3         3.0         0.2   setosa
## ---
## 148:    7.7          6.7         2.8         2.0 virginica
## 149:    7.7          6.9         2.6         2.3 virginica
## 150:    7.9          6.4         3.8         2.0 virginica
```

As with `setnames()`, `setcolorder()` may be used with numeric positions rather than variable names. We mention here as well that these are all done via reference changes, and our data table `diris` remains keyed throughout all of this:

```
v <- c(1, 3, 2, 4, 5)
setcolorder(diris, v)
diris
```

```
##      SepalL Sepal.Width Petal.Length Petal.Width  Species
##   1:    4.3         3.0          1.1         0.1   setosa
##   2:    4.4         3.2          1.3         0.2   setosa
##   3:    4.4         3.0          1.3         0.2   setosa
## ---
## 148:    7.7         2.8          6.7         2.0 virginica
## 149:    7.7         2.6          6.9         2.3 virginica
## 150:    7.9         3.8          6.4         2.0 virginica
```

7.4 Creating Variables

We next turn our attention to creating new variables within the data table. Thinking back to a for loop structure, the goal would be to step through each row of our data table and make some analysis or change. A data table has very efficient code to do precisely this without formally calling a for loop or even using the *apply() functions.

By efficient, we mean faster than either of those two methods in all, or almost all, cases. To get us started on solid ground, we go ahead and re-create and rekey our data table based on the original iris data set to remove any changes we made in the prior sections:

```
diris <- as.data.table(iris)
setkey(diris, Species)
```

Note   New variables are created in a data table by using the `:=` operator in the `j` formal, `data[, newvar := value]`. Variables from the data set can be used as well as using functions.

Creating a new variable in a data table is creating a new column. Because of that, we operate in the second formal, or `j`, area of our generic structure `Data.Table[i, j, by]`. Column creation involves using the assignment by reference `:=` operator. Column creation can take different forms depending on what we wish to create. We may create just a single column and populate it with zeros, create multiple columns with different variables, or even create a new column based on calculations on existing columns. Notice that all these are in the second formal argument location in our data table call, and pay special attention to the way we create multiple columns at once, named X1 and X2, by using `.()` to pass a vector of column names on the left and a list of column values on the right-hand side:

```
diris[, V0 := 0]
diris[, Sepal.Length := NULL]
diris[, c("X1", "X2") := .(1L, 2L)]
diris[, V := Petal.Length * Petal.Width]
diris
```

```
##      Sepal.Width Petal.Length Petal.Width    Species V0 X1 X2     V
## 1:          3.5          1.4         0.2     setosa  0  1  2  0.28
## 2:          3.0          1.4         0.2     setosa  0  1  2  0.28
## 3:          3.2          1.3         0.2     setosa  0  1  2  0.26
## ---
## 148:        3.0          5.2         2.0 virginica  0  1  2 10.40
## 149:        3.4          5.4         2.3 virginica  0  1  2 12.42
## 150:        3.0          5.1         1.8 virginica  0  1  2  9.18
```

Assignment to NULL deletes columns. Full deletion in this way removes any practical ability to restore that column, but it is fast. Another option, of course, would be to use - to generate a subset of the original data table and then copy that to a second data table. The difference is in speed and memory usage. Making a copy may almost double your memory usage and takes linear time to copy based on total elements. Of course, making a copy keeps the original around if you want or need to recover all original data. It is possible to delete more than one column at a time via assignment to NULL. We show just the first row to verify the deletion was successful:

```
diris[, c("V", "V0") := NULL]
diris[1]
```

```
##    Sepal.Width Petal.Length Petal.Width Species X1 X2
## 1:         3.5          1.4         0.2  setosa  1  2
```

It is of great benefit to be able to generate new columns based on computations involving current columns. Additionally, this process can be readily modified by commands in the first formal area that allows us to select key data results. Recall that our key is of species, one of which is setosa. We re-create our deleted V column with the multiplication of two current columns and do so only for rows with the setosa Species key. Notice that rows not matching the key get NA as filler:

```
diris["setosa", V := Petal.Length * Petal.Width]
unique(diris)
```

```
##      Sepal.Width Petal.Length Petal.Width  Species X1 X2    V
##   1:         3.5          1.4         0.2   setosa  1  2 0.28
##   2:         3.0          1.4         0.2   setosa  1  2 0.28
##   3:         3.2          1.3         0.2   setosa  1  2 0.26
## ---
## 141:         3.0          5.2         2.0 virginica  1  2   NA
## 142:         3.4          5.4         2.3 virginica  1  2   NA
## 143:         3.0          5.1         1.8 virginica  1  2   NA
```

However, if we later decide that virginica also deserves some values in the V column, that can be done, and it need not involve the same calculation as used for setosa. We again use the unique() function call to look only at the first example of each species. While this example is artificial, imagine a distance column being created based on coordinates already stored in the data. Based on key values, perhaps some distance calculations ought to be in Manhattan length, while others might naturally be in Euclidean length:

```
diris["virginica", V := sqrt(Petal.Length * Petal.Width)]
unique(diris)
```

```
##        Sepal.Width Petal.Length Petal.Width    Species X1 X2    V
##    1:          3.5          1.4         0.2     setosa  1  2 0.28
##    2:          3.0          1.4         0.2     setosa  1  2 0.28
##    3:          3.2          1.3         0.2     setosa  1  2 0.26
## ---
## 141:          3.0          5.2         2.0  virginica  1  2 3.22
## 142:          3.4          5.4         2.3  virginica  1  2 3.52
## 143:          3.0          5.1         1.8  virginica  1  2 3.03
```

Such calculations are not limited to column creation. Suppose we want to know the mean of all values in Sepal.Width. That calculation can be performed in several ways, each of which may be advantageous depending on your final goal. Recall data tables are data frames as well; we could call our arithmetic mean function on a suitable data-frame-style call. However, because we want to find the arithmetic mean of the values of a particular column, we could also call it from the second formal. Remember when we mentioned caution was required for items in this j location? It is possible to access values from j as vectors. Contrastingly, should we wish to return a data table, that may too be done. All three of these return the same mean, after all. However, it is the last that may be the most instructive:

```
mean(diris$Sepal.Width)
```

```
## [1] 3.1
```

```
diris[, mean(Sepal.Width)]
```

```
## [1] 3.1
```

```
diris[, .(M = mean(Sepal.Width))]
```

```
##        M
## 1: 3.1
```

The last line is perhaps best, because it can readily expand. Suppose we want to know the arithmetic mean as in the preceding code, but segregated across our species. This would call for using both the second and third formals. Suppose we also want to know more than just one column's mean. Well, the .() notation would allow us to have a convenient way to store that new information in a data table format, thus making it ready to be used for analysis. Notice we now have the intraspecies means, and in the second bit of our code, we have it for more than one column, and in both cases, the returned results are data tables. Thus, we could access information from these new constructs just as we have been doing all along with diris. Much like the hole Alice falls down to reach Wonderland, it becomes a question of "How deep do you want to go?" rather than "How deep can we go?"

```
diris[, .(M = mean(Sepal.Width)), by = Species]

##         Species   M
## 1:       setosa 3.4
## 2: versicolor 2.8
## 3:  virginica 3.0

diris[, .(M1 = mean(Sepal.Width),
          M2 = mean(Petal.Width)), by = Species]

##         Species  M1    M2
## 1:       setosa 3.4 0.25
## 2: versicolor 2.8 1.33
## 3:  virginica 3.0 2.03
```

To see the full power of these new data tables we are building, suppose we want to see whether there is any correlation between the means of species' sepal and petal widths. Further imagine that we might find a use for the mean data at some future point beyond the correlation. We can treat the second data table in the preceding code snippet as a data table (because it is) and use the second formal to run correlation. This code follows a Data.Table[i, j, by] [i, j, by] layout, which sounds rather wordy written out; take a look at the following code to see how it works:

```
diris[, .(M1 = mean(Sepal.Width),
          M2 = mean(Petal.Width)), by = Species][,
  .(r = cor(M1, M2))]

##           r
## 1: -0.76
```

Variables may create variables. In other words, a new variable may be created that indicates whether a petal width is greater or smaller than the median petal width for that particular species. In the code that follows, median(Petal.Width) is calculated by Species, and the boolean values are stored in the newly created column MedPW for all 150 rows:

```
diris[, MedPW := Petal.Width > median(Petal.Width), by = Species]
diris
```

```
##      Sepal.Width Petal.Length Petal.Width    Species X1 X2     V MedPW
## 1:           3.5          1.4         0.2     setosa  1  2 0.28 FALSE
## 2:           3.0          1.4         0.2     setosa  1  2 0.28 FALSE
## 3:           3.2          1.3         0.2     setosa  1  2 0.26 FALSE
## ---
## 148:         3.0          5.2         2.0  virginica  1  2 3.22 FALSE
## 149:         3.4          5.4         2.3  virginica  1  2 3.52  TRUE
## 150:         3.0          5.1         1.8  virginica  1  2 3.03 FALSE
```

Just as it was possible to have multiple arguments passed to the second formal, it is possible to have multiple arguments passed to the third via the same . () list function call. Again, because what is returned is a data table, we treat it as one with the brackets and perform another layer of calculations on our results:

```
diris[, .(M1 = mean(Sepal.Width),
          M2 = mean(Petal.Width)),
       by = .(Species, MedPW)]
```

```
##        Species MedPW  M1   M2
## 1:      setosa FALSE 3.4 0.19
## 2:      setosa  TRUE 3.5 0.37
## 3:  versicolor  TRUE 3.0 1.50
## 4:  versicolor FALSE 2.6 1.19
## 5:   virginica  TRUE 3.1 2.27
## 6:   virginica FALSE 2.8 1.81
```

```
diris[, .(M1 = mean(Sepal.Width),
          M2 = mean(Petal.Width)),
       by = .(Species, MedPW)][,
   .(r = cor(M1, M2)), by = MedPW]
```

```
##      MedPW       r
## 1: FALSE -0.78
## 2:  TRUE -0.76
```

Performing computations in the j argument is one of the most powerful features of data tables. For example, we have shown relatively simple computations of means, medians, and correlations. However, nearly any function can be used. The only caveat is that if you want the output also to be a data table, the function should return or be manipulated to return output appropriate for a data table. For instance, a linear model object that is a list of varying length elements would be messy to coerce into a data table, as it is not tabular data. However, you could easily extract regression coefficients as a vector, and that could become a data table.

7.5 Merging and Reshaping Data

Managing data often involves a desire to combine data from different sources into a single object or to reshape data in one object to a different format. When combining data, there are three distinct types of data merge: one-to-one, one-to-many (or many-to-one), and many-to-many. These merges are introduced in this section. The general process is to look at data from two or more collections and then compare keys, with a goal to successfully connect information by row. This is called horizontal merging because it increases the horizontal length of the target data set.

Merging Data

Merges in which each key is unique per data set are of the one-to-one form. An example is merging data sets keyed by government Social Security number (SSN) that contain legal residence addresses in one and preferred emails in another. In that case, we would expect the real-world data set to have one address per SSN, and the preferred email data set to have only one row per SSN. Contrastingly, a one-to-many or many-to-one merge might occur if we had our official residence address data set as well as all known email addresses, where each new email address tied to one SSN had its row. If the goal were to append all emails to a single real-world address row, it would be many-to-one, and column names such as Email001 and Email002 might collapse to a single row. On the other hand, if the goal were to append a real-world address to each email, then perhaps only one new column would be added, although the data copies to each row of the email address. Finally, there can be many-to many-merges, in which multiple rows of data append to multiple rows.

It is worth noting the main difficulty with merging may well be in understanding what is being done rather than understanding the code itself. For data tables, the function call is merge(), and arguments control the type of merge performed. Due to the keyed nature of data tables, merging by the key choice can be quite efficient. Choosing to merge by non-key columns is also possible, yet will not be as efficient computationally.

Before getting into the first merge code example, we generate four data tables with information unique to each. All are keyed by Species from the iris data set. In real life, such code is almost never required because normally data tables come already separated. This code is included only for reproducibility of the merge results. Also note these are very short data sets; the goal here is for them to be viewed and fully understood independently. This provides the best environment from which to understand the final merge results:

```
diris <- diris[, .(Sepal.Width = Sepal.Width[1:3]), keyby = Species]
diris2 <- unique(diris)
dalt1 <- data.table(
   Species = c("setosa", "setosa", "versicolor", "versicolor",
               "virginica", "virginica", "other", "other"),
   Type = c("wide", "wide", "wide", "wide",
            "narrow", "narrow", "moderate", "moderate"),
  MedPW = c(TRUE, FALSE, TRUE, FALSE, TRUE, FALSE, TRUE, FALSE))
setkey(dalt1, Species)

dalt2 <- unique(dalt1)[1:2]

diris

##        Species Sepal.Width
## 1:      setosa         3.5
## 2:      setosa         3.0
## 3:      setosa         3.2
## 4: versicolor         3.2
## 5: versicolor         3.2
## 6: versicolor         3.1
```

```
## 7:  virginica        3.3
## 8:  virginica        2.7
## 9:  virginica        3.0
```

diris2

```
##          Species Sepal.Width
## 1:        setosa         3.5
## 2:        setosa         3.0
## 3:        setosa         3.2
## 4: versicolor         3.2
## 5: versicolor         3.1
## 6:  virginica         3.3
## 7:  virginica         2.7
## 8:  virginica         3.0
```

dalt1

```
##          Species       Type MedPW
## 1:         other moderate  TRUE
## 2:         other moderate FALSE
## 3:        setosa     wide  TRUE
## 4:        setosa     wide FALSE
## 5: versicolor     wide  TRUE
## 6: versicolor     wide FALSE
## 7:  virginica   narrow  TRUE
## 8:  virginica   narrow FALSE
```

dalt2

```
##    Species     Type MedPW
## 1:   other moderate  TRUE
## 2:   other moderate FALSE
```

Note A one-to-one merge requires two keyed data tables, with each row having a unique key: `merge(x = dataset1, y = dataset2)`. The merge exactly matches a row from one data table to the other, and the result includes all the columns (variables) from both data tables. Whether nonmatching keys are included is controlled by the all argument, which defaults to `FALSE` to include only matching keys, but may be set to `TRUE` to include nonmatching keys as well.

In a one-to-one merge, there is at most one row per key. In the case of `diris2`, there are three rows, and each has a unique species. Similarly, in `dalt2`, there are two rows, and each has a unique species. The only common key between these two data sets is `setosa`. A `merge()` call on this one-to-one data compares keys and by default creates a new data table combining only those rows with matching keys. In this case, merging yields a tiny data table with one row. The original two data tables are not changed in this operation. The call to `merge()` takes a minimum of two arguments; the first is x, and the second is y, which represent the two data sets to be merged:

```
merge(diris2, dalt2)

## Empty data.table (0 rows and 4 cols): Species,Sepal.Width,Type,MedPW
```

Notice this process creates a complete data set. However, it is quite small, because the only results returned were those for which full and complete matches were possible. Depending on the situation, different results may be desired from a one-to-one merge. Perhaps all rows of `diris2` are wanted regardless of whether they match `dalt2`. Conversely, the opposite may be the case, and all rows of `dalt2` may be desired regardless of whether they match `diris2`. Finally, to get all possible data, it may be desirable to keep all rows that appear in either data set. These objectives may be accomplished by variations of the `all = TRUE` formal argument. In particular, note how to control which single data set is designated as essential to keep via the `.x` or `.y` suffix to `all`. If data is not available for a specific row or variable, the cell is filled with a missing value (NA):

```
merge(diris2, dalt2, all.x = TRUE)

##          Species Sepal.Width Type MedPW
## 1:        setosa         3.5 <NA>    NA
## 2:        setosa         3.0 <NA>    NA
## 3:        setosa         3.2 <NA>    NA
```

```
## 4: versicolor          3.2 <NA>    NA
## 5: versicolor          3.1 <NA>    NA
## 6:  virginica          3.3 <NA>    NA
## 7:  virginica          2.7 <NA>    NA
## 8:  virginica          3.0 <NA>    NA
```

merge(diris2, dalt2, **all**.y = TRUE)

```
##     Species Sepal.Width     Type MedPW
## 1:    other          NA moderate  TRUE
## 2:    other          NA moderate FALSE
```

merge(diris2, dalt2, **all** = TRUE)

```
##           Species Sepal.Width     Type MedPW
##  1:        other          NA moderate  TRUE
##  2:        other          NA moderate FALSE
##  3:       setosa         3.5    <NA>    NA
##  4:       setosa         3.0    <NA>    NA
##  5:       setosa         3.2    <NA>    NA
##  6: versicolor         3.2    <NA>    NA
##  7: versicolor         3.1    <NA>    NA
##  8:  virginica         3.3    <NA>    NA
##  9:  virginica         2.7    <NA>    NA
## 10:  virginica         3.0    <NA>    NA
```

Note A one-to-many or many-to-one merge involves one data table in which each row has a unique key and another data table in which multiple rows may share the same key. Rows from the data table with unique keys are repeated as needed to match the data table with repeated keys. In R, identical code is used as for a one-to-one merge: merge(x = dataset1, y = dataset2).

Merges involving diris and dalt2 are of the one-to-many or many-to-one variety. Again, the behavior wanted in the returned results may be controlled when it comes to keeping only full matches or allowing various flavors of incomplete data to persist through the merge. The three examples provided are not exhaustive, yet they are representative of the different possibilities. The last two cases may not be desirable, as other has only one row, whereas most species have three:

```
merge(diris, dalt2)
```

```
## Empty data.table (0 rows and 4 cols): Species,Sepal.Width,Type,MedPW
```

```
merge(diris, dalt2, all.y = TRUE)
```

```
##      Species Sepal.Width     Type MedPW
## 1:    other          NA moderate  TRUE
## 2:    other          NA moderate FALSE
```

```
merge(diris, dalt2, all = TRUE)
```

```
##           Species Sepal.Width     Type MedPW
##  1:         other          NA moderate  TRUE
##  2:         other          NA moderate FALSE
##  3:        setosa         3.5     <NA>    NA
##  4:        setosa         3.0     <NA>    NA
##  5:        setosa         3.2     <NA>    NA
##  6:    versicolor         3.2     <NA>    NA
##  7:    versicolor         3.2     <NA>    NA
##  8:    versicolor         3.1     <NA>    NA
##  9:     virginica         3.3     <NA>    NA
## 10:     virginica         2.7     <NA>    NA
## 11:     virginica         3.0     <NA>    NA
```

So far, one-to-one and one-to-many merges have been demonstrated. These types of merges, or joins, have a final, merged number of rows that is at most the sum of the rows of the two data tables. As long as at least one of the dimensions of the merge is capped to one, there is no concern. However, in a many-to-many scenario, the upper limit of size becomes the product of the row count of the two data tables. In many research contexts, many-to-many merges are comparatively rare. Thus, the default behavior of a merge for data tables is to prevent such a result, and `allow.cartesian = FALSE` is the mechanism through which this is enforced as a formal argument of `merge()`. The same is not true for other data structures but is unique to data tables.

Attempting to merge `diris` and `dalt1` results in errors from this default setting, which disallows merged table rows to be larger than the total sum of the rows. With three setosa keys in `diris` and two setosa keys in `dalt1`, a merged table would result with 3 x 2 = 6 total rows for setosa alone. As many-to-many merges involve a multiplication of the total number of rows, it is easy to end up with a very large database. Allowing this would yield the following results. Notice that this result has only complete rows:

```
merge(diris, dalt1, allow.cartesian = TRUE)
```

```
##           Species Sepal.Width   Type MedPW
##  1:        setosa         3.5   wide  TRUE
##  2:        setosa         3.5   wide FALSE
##  3:        setosa         3.0   wide  TRUE
##  4:        setosa         3.0   wide FALSE
##  5:        setosa         3.2   wide  TRUE
##  6:        setosa         3.2   wide FALSE
##  7:    versicolor         3.2   wide  TRUE
##  8:    versicolor         3.2   wide FALSE
##  9:    versicolor         3.2   wide  TRUE
## 10:    versicolor         3.2   wide FALSE
## 11:    versicolor         3.1   wide  TRUE
## 12:    versicolor         3.1   wide FALSE
## 13:     virginica         3.3 narrow  TRUE
## 14:     virginica         3.3 narrow FALSE
## 15:     virginica         2.7 narrow  TRUE
## 16:     virginica         2.7 narrow FALSE
## 17:     virginica         3.0 narrow  TRUE
## 18:     virginica         3.0 narrow FALSE
```

The preceding has only complete rows and is missing the other species. Allowing all possible combinations nets two more rows, for a total of 20. It is an entirely different conversation if the resulting data is useful. When merging data, it always pays to consider what types of information are available and what might be needed for a particular bit of analysis:

```
merge(diris, dalt1, all = TRUE, allow.cartesian = TRUE)
```

```
##           Species Sepal.Width     Type MedPW
##  1:         other          NA moderate  TRUE
##  2:         other          NA moderate FALSE
##  3:        setosa         3.5     wide  TRUE
##  4:        setosa         3.5     wide FALSE
##  5:        setosa         3.0     wide  TRUE
##  6:        setosa         3.0     wide FALSE
##  7:        setosa         3.2     wide  TRUE
##  8:        setosa         3.2     wide FALSE
##  9:    versicolor         3.2     wide  TRUE
## 10:    versicolor         3.2     wide FALSE
## 11:    versicolor         3.2     wide  TRUE
## 12:    versicolor         3.2     wide FALSE
## 13:    versicolor         3.1     wide  TRUE
## 14:    versicolor         3.1     wide FALSE
## 15:     virginica         3.3   narrow  TRUE
## 16:     virginica         3.3   narrow FALSE
## 17:     virginica         2.7   narrow  TRUE
## 18:     virginica         2.7   narrow FALSE
## 19:     virginica         3.0   narrow  TRUE
## 20:     virginica         3.0   narrow FALSE
```

For all of these, the key is Species, and that is only one column. Furthermore, it is nonunique. It is also possible to have keys based on multiple columns. Recognize that key choice may make all the difference between a one-to-one and a many-to-many merge. Consider the following instructive example, in which only a single column is used as the key and the data is merged:

```
d2key1 <- data.table(
  ID1 = c(1, 1, 2, 2),
  ID2 = c(1, 2, 1, 2),
  X = letters[1:4])
```

```
d2key2 <- data.table(
  ID1 = c(1, 1, 2, 2),
  ID2 = c(1, 2, 1, 2),
  Y = LETTERS[1:4])
```

```
d2key1
```

```
##    ID1 ID2 X
## 1:   1   1 a
## 2:   1   2 b
## 3:   2   1 c
## 4:   2   2 d
```

```
d2key2
```

```
##    ID1 ID2 Y
## 1:   1   1 A
## 2:   1   2 B
## 3:   2   1 C
## 4:   2   2 D
```

```
setkey(d2key1, ID1)
setkey(d2key2, ID1)
```

```
merge(d2key1, d2key2)
```

```
##    ID1 ID2.x X ID2.y Y
## 1:   1     1 a     1 A
## 2:   1     1 a     2 B
## 3:   1     2 b     1 A
## 4:   1     2 b     2 B
## 5:   2     1 c     1 C
## 6:   2     1 c     2 D
## 7:   2     2 d     1 C
## 8:   2     2 d     2 D
```

For the merge to be successful with its duplicated column names, ID2.x and ID2.y were created. The fact is this data should have resulted in only four rows, once merged, because the data in those two columns is identical! By setting the key for both data sets to involve both columns ID1 and ID2, not only is the many-to-many merge reduced to a one-to-one merge but the data is of a more practical nature. Multiple columns being used to key can be especially useful when merging multiple long data sets. For instance, with repeated measures on multiple people over time, there could be an ID column for participants and another time variable, but both would be used to merge properly:

```
setkey(d2key1, ID1, ID2)
setkey(d2key2, ID1, ID2)

merge(d2key1, d2key2)

##      ID1 ID2 X Y
## 1:    1   1 a A
## 2:    1   2 b B
## 3:    2   1 c C
## 4:    2   2 d D
```

Reshaping Data

In addition to being merged, data may be reshaped. Consider our iris data set: each row represents a unique flower measurement, yet many columns are used for each flower measured. Despite not having a unique key, there are a variety of measurements for each flower (row). Wider data may be reshaped to long data, in which each key appears multiple times and stores variables in fewer columns. First, we reset our data to the iris data set and create a unique ID to be our key based on the row position. Recall data tables suppress rows, so we add those into their ID column by using another useful shorthand of data tables, namely, .N, for number of rows:

```
diris <- as.data.table(iris)

diris[, ID := 1:.N]
setkey(diris, ID)
```

Note   Reshaping is used to transform data from wider to longer or from longer to wider format. In wide format, a repeatedly measured variable has separate columns, with the column name indicating the assessment. In long format, a repeated variable has only one column, and a second column indicates the assessment or "time." Wide-to-long reshaping is accomplished by using the `melt()` function, which requires several arguments.

To reshape long, we `melt()` the data table measure variables into a stacked format. This function takes several arguments, which are discussed in order. To see all the arguments, examine the help `?melt.data.table`. It is important to specify the exact method, `melt.data.table`, because the arguments for the generic function are different.

The first formal, `data`, is the name of the data table to be melted, followed by `id. vars`, which is where the ID variables and any other variable that should not be reshaped long belong. By default, `id.vars` is any column name not explicitly called out in the next formal. Any column name assigned to `id.vars` is not reshaped into long form; it is repeated, but left as is. The third formal is for the measure variables, and `measure. vars` uses these column names to stack the data. Typically, this is a list with separate character vectors to indicate which variables should be stacked together in the long data. The fourth formal is `variable.name`, which is the name of the new variable created to indicate which level of the `measure.vars` a specific row in the new long data set belongs to. In the following example, the first level is for length measurements, while the second level corresponds to width measurements of iris flowers. Finally, `value.name` is the name of the molten (or long) column(s), which is (are) the names of the columns in our long data containing the original wide variables' values.

Describing reshaping and the required arguments is rather complex. It may be easier to compare output between the now familiar `iris` data set and the reshaped, long data set, shown here:

```
diris.long <- melt(diris,
  measure.vars = list(
    c("Sepal.Length", "Sepal.Width"),
    c("Petal.Length", "Petal.Width")),
  variable.name = "Type",
  value.name = c("Sepal", "Petal"),
  id.vars = c("ID", "Species"))
diris.long
```

```
##          ID   Species Type Sepal Petal
##     1:    1    setosa    1   5.1   1.4
##     2:    2    setosa    1   4.9   1.4
##     3:    3    setosa    1   4.7   1.3
## ---
## 298: 148 virginica    2   3.0   2.0
## 299: 149 virginica    2   3.4   2.3
## 300: 150 virginica    2   3.0   1.8
```

Notice our data now takes 300 rows because of being melted. The Type 1 vs. 2 may be confusing to equate with length vs. width measurements for sepals and petals. The melt() method for data tables defaults to just creating an integer value indicating whether it is the first, second, or third (and so forth) level. To make the labels more informative, we use factor(). Because factor will operate on a column and change the row elements of the column, it belongs in the second formal j argument of the data table:

```
diris.long[, Type := factor(Type, levels = 1:2, labels = c("Length",
"Width"))]
diris.long
```

```
##          ID   Species   Type Sepal Petal
##     1:    1    setosa Length   5.1   1.4
##     2:    2    setosa Length   4.9   1.4
##     3:    3    setosa Length   4.7   1.3
## ---
## 298: 148 virginica  Width   3.0   2.0
## 299: 149 virginica  Width   3.4   2.3
## 300: 150 virginica  Width   3.0   1.8
```

The data set diris.long is a good example of a long data set. It is not, however, the longest possible data set from diris—not that such a thing is necessarily a goal, by any means. However, should melt() be called on just diris along with calling out id.vars, the resulting data table is perhaps the longest possible form of the iris data. Just as id.vars defaults to any values not in measure.vars, so too measure.vars defaults to any values not in id.vars. As you can also see, variable.name and value.name default to variable and value, respectively. Notice that diris.long2 has 600 rows, and what were formerly the four columns of sepal and petal lengths and widths are now two columns. Also note that when this is done, melt() automatically sets the values of variable to the initial variable names from the data set:

```
diris.long2 <- melt(diris, id.vars = c("ID", "Species"))
diris.long2
```

```
##           ID   Species      variable value
##    1:    1    setosa Sepal.Length   5.1
##    2:    2    setosa Sepal.Length   4.9
##    3:    3    setosa Sepal.Length   4.7
## ---
## 598: 148 virginica  Petal.Width   2.0
## 599: 149 virginica  Petal.Width   2.3
## 600: 150 virginica  Petal.Width   1.8
```

Note To reshape data from long to wide, the dcast() function is used. Two required arguments are data and formula, which indicate how the data should be reshaped from long to wide. The formula must indicate unique values in a cell, or the data needs to be aggregated within a cell.

While melt() is used to make wide data long, dcast() is used to make long data wide. Without unique identifiers for data, multiple values must be aggregated to fit in one cell. Thus, calling our casting function without some thought may create unwanted results (although sometimes aggregating multiple values is exactly what is desired). Notice that in the second formal, Species is called out as the identity key, and variable is named as the location to find our new column names:

```
dcast(diris.long2, Species ~ variable)
```

```
## Aggregate function missing, defaulting to 'length'
```

```
##        Species Sepal.Length Sepal.Width Petal.Length Petal.Width
## 1:      setosa           50          50           50          50
## 2: versicolor           50          50           50          50
## 3:  virginica           50          50           50          50
```

If this long data is to be better cast to wide, a unique identifier is needed so the formula creates unique cells that contain only one value. Unless aggregating results is desired, for a given variable, the formula argument should result in a single value, not multiple values.

In this case, since there are four possible values that the variable column takes on in the long data, as long as there is a unique key that is never duplicated over a specific variable such as Sepal.Length, the cast works out. In other words, if the wide data has a unique key, we can combine that unique key with the variable names in the long data, which together will exactly map cells from the long data set back to the wide data. The left side of the formula contains the key or ID variable. This side of the formula (left of the tilde, ˜) may also contain any other variable that does not vary (such as would have been specified using the id.vars argument to melt() if the data had initially been reshaped from wide to long). The right-hand side of the formula (after the tilde, ˜) contains the variable or variables that indicate which row of the long data set belongs in which column of the new wide data set.

In our long iris data set example, the ID and Species variables do not vary and so belong on the left side. The variable called variable indicates whether a particular value in the long data set represents Sepal.Length, Sepal.Width, Petal.Length, or Petal. Width. Thus, for this example, the formula looks like ID + Species ˜variable:

```
diris.wide2 <- dcast(diris.long2, ID + Species ˜ variable)
diris.wide2
```

##	ID	Species	Sepal.Length	Sepal.Width	Petal.Length	Petal.Width
## 1:	1	setosa	5.1	3.5	1.4	0.2
## 2:	2	setosa	4.9	3.0	1.4	0.2
## 3:	3	setosa	4.7	3.2	1.3	0.2
## ---						
## 148:	148	virginica	6.5	3.0	5.2	2.0
## 149:	149	virginica	6.2	3.4	5.4	2.3
## 150:	150	virginica	5.9	3.0	5.1	1.8

Going back to the first diris.long data with the two-level factor for length and width of sepals and petals, it is also possible to cast this back to wide. The call to dcast() begins similarly to our previous example. We specify the long data set, a formula indicating the ID and Species variables that are not stacked, and the variable, Type, that indicates what each row of the data set belongs to.

In such not-so-long (for lack of a better word) data, we require an additional argument, the `value.var` formal, to let `dcast()` know which variables in the long data are repeated measures. This can be a character vector of every variable in the long data set that will correspond to multiple variables in the wide data set. In our case, these are Sepal and Petal, which in the long data contain both lengths and widths, as indicated in the Type variable. Additionally, the separator must be chosen, as data table column names may have different separators. This is selected via the `sep =` formal, and here we use the full stop for continuity's sake, as that is how the variables were labeled in the original wide iris data set. We show merely the `dcast()` command and check that regardless of method, the wide data is recovered via the `all.equal()` function, which checks for precisely what it states. To view the results, type `diris.wide` at the console:

```
diris.wide <- dcast(diris.long, ID + Species ~ Type,
      value.var = list("Sepal", "Petal"),
      sep = ".")
```

```
all.equal(diris.wide2, diris.wide)
```

```
## [1] TRUE
```

7.6 Summary

In this chapter, you met one of the two powerful (and more modern) ways of managing data in R. In particular, the type of data we manipulate with `data.table` is column and row, or table, data. In the next chapter, we delve more deeply into data manipulation. From there, you will meet `dplyr` [29], which is the other modern way of managing data in R. We will close out this section with a look at databases outside R. Table 7-1 provides a summary of key functions used in this chapter.

Table 7-1. *Listing of key functions described in this chapter and summary of what they do*

Function	What It Does
d[i, j, by]	General structure of a call to subset, compute in, or modify a data table.
as.data.table()	Coerces a data structure to a data table.
setkey()	Takes a data table in the first formal and then creates a key based on the column name arguments next passed to the function.
haskey()	This function is called on a data table and returns a boolean value.
key()	Returns the data table's key if it exists.
order()	Controls the order to be ascending or descending for the data of a column.
tables()	Returns any current data tables currently active.
anyDuplicated()	Returns the location of the first duplicated entry.
duplicated()	Returns TRUE or FALSE for all rows to show duplication. Can take a key. Otherwise, it is by all columns.
unique()	Returns a data table that has only unique rows. Again, these may be specified to be specific rows. If there is a key, it is based on the key. If there is no key, it is based on all columns.
setnames()	Allows column names to be changed.
setcolorder()	Allows column order to be adjusted.
:=	Assignment operator within a data table.
.()	Shorthand for list().
.N	Shorthand for number of rows in a data table or number of rows in a subset, such as when a compute operation is performed by a grouping variable.
merge()	Merges two data tables based on matching key columns.
melt()	Converts wide data to long data.
dcast()	Converts long data to wide data.

CHAPTER 8

Advanced `data.table`

We already introduced the `data.table` package which is the heart of this chapter, covering the basics of accessing, editing, and manipulating data under the broad term data management. Although not glamorous, data management is a critical first step to data visualization or analysis. Furthermore, the majority of time on a particular analysis project often comes from data management. For example, running a linear model in R takes one line of code, once the data is clean and in the expected format. Data management is challenging because raw data comes in all types, shapes, and formats and missing data is common. In addition, you may also have to combine or merge separate data sources. In this chapter, we go beyond the basic use of `data.table` to more complex data management tasks.

There tend to be two flavors of data wrangling. One-time conversions are often manual, as writing code often is not efficient if it is not reused (e.g., changing one or two variable names or renaming a data file to be consistent). On the other hand, for operations needing repetition (e.g., renaming or labeling hundreds of variables) or working with larger data, more programming is used for data management. Install (if necessary) the `stringdist` package [23]; another needed library is `foreign` [15]:

```
library(stringdist)
library(data.table)
library(foreign)

options(stringsAsFactors = FALSE,
        datatable.print.nrows = 20, ## if over 20 rows
        datatable.print.topn = 3, ## print first and last 3
        digits = 2) ## reduce digits printed by R
```

© Matt Wiley and Joshua F. Wiley 2020
M. Wiley and J. F. Wiley, *Advanced R 4 Data Programming and the Cloud*,
https://doi.org/10.1007/978-1-4842-5973-3_8

8.1 Data Munging/Cleaning

To obtain data requiring the *programming* flavor of munging, we download data from the National Survey of Children's Health, 2003 [4], at https://doi.org/10.3886/ ICPSR04691.v1, where we chose the *Stata* format. We place the folder in our data/ ch08/ folder inside our working directory and use read.dta() to input the data to R. After converting it to a data table, we set the key to be the identification number column IDNUMR:

```
d <- read.dta("data/ch08/ICPSR_04691/DS0001/04691-0001-Data.dta")
d <- as.data.table(d)
setkey(d, IDNUMR)
```

To get a sense of the structure of this much data (there are over 300 variables), we use the str() function. We suppress attributes in order to focus on the column names and the types of data inside each column. Setting strict.width = "cut" enforces the options we set earlier, and we look at only the first 20 columns:

```
str(d, give.attr = FALSE, strict.width = "cut", list.len = 20)
## Classes 'data.table' and 'data.frame':      102353 obs. of 301 variables:
##  $ IDNUMR  : int  1 2 3 4 5 6 7 8 9 10 ...
##  $ STATE   : Factor w/ 51 levels "1-AK","2-AL",..: 35 25 45 18 36 37..
##  $ MSA_STAT: Factor w/ 4 levels "-2 - MISSING",..: 4 4 4 4 4 4 4 4 4..
##  $ AGEYR_CH: int  12 1 10 7 9 11 9 15 17 1 ...
##  $ TOTKIDS4: Factor w/ 4 levels "1 - 1 CHILD",..: 1 1 1 3 4 2 3 3 1 ..
##  $ AGEPOS4 : Factor w/ 5 levels "1 - ONLY CHILD",..: 1 1 1 4 2 2 3 2..
##  $ S1Q01   : Factor w/ 8 levels "-4 - PARTIAL INTERVIEW",..: 6 5 6 5..
##  $ RELATION: Factor w/ 6 levels "-2 - MISSING",..: 3 2 3 3 2 2 2 4 3..
##  $ TOTADULT: Factor w/ 6 levels "-2 - MISSING",..: 4 3 2 3 2 3 3 2 4..
##  $ EDUCATIO: Factor w/ 6 levels "-2 - MISSING",..: 4 4 4 4 3 4 4 3 4..
##  $ PLANGUAG: Factor w/ 5 levels "-2 - MISSING",..: 2 2 2 2 2 2 2 2 2..
##  $ S2Q01   : Factor w/ 11 levels "-4 - PARTIAL INTERVIEW",..: 5 5 5 ..
##  $ S2Q02R  : int  59 29 49 45 96 64 51 64 65 33 ...
##  $ HGHT_FLG: int  0 0 0 0 0 0 0 0 0 0 ...
##  $ S2Q03R  : int  100 20 55 60 98 115 64 135 115 26 ...
##  $ WGHT_FLG: int 0 0 0 0 0 0 0 0 0 0 ...
##  $ BMICLASS: Factor w/ 5 levels "-2 - MISSING",..: 3 1 3 5 1 3 3 3 3..
```

```
## $ S2Q04    : Factor w/ 8 levels "-4 - PARTIAL INTERVIEW",..: 5 5 5 5..
## $ S2Q05    : Factor w/ 8 levels "-4 - PARTIAL INTERVIEW",..: 4 4 4 4..
## $ S2Q06    : Factor w/ 8 levels "-4 - PARTIAL INTERVIEW",..: 4 4 4 4..
##   [list output truncated]
```

Many columns contain missing data that is currently stored as levels of a factor, rather than as R's NA to indicate to R those values are missing. With over 100,000 rows, we definitely need to automate any changes. From the study's documentation, we find the values used that we must recode to missing. These values have consistent labels, but their numbers change depending on the number of legitimate levels in a variable. We use the table() function to generate a frequency table of the unique values in a variable. This is useful both as a way to see the unique values and to become familiar with the data and how much is missing (relatively little for variables such as *education* and more for BMI class). We want to recode partial *interview, not in universe, missing, legitimate skip, don't know*, and *refused* to NA. However, the labels are not identical; EDUCATIO has the number 97 for refused, whereas S1Q01 has 7 for refused:

```
table(d[, EDUCATIO])
##
##                      -2 - MISSING          1 - LESS THAN HIGH SCHOOL
##                                 3                               4661
## 2 - 12 YEARS, HIGH SCHOOL GRADUATE          3 - MORE THAN HIGH SCHOOL
##                             21238                              76022
##                 96 - DON'T KNOW                        97 - REFUSED
##                               324                                105

table(d[, PLANGUAG])

##
##          -2 - MISSING         1 - ENGLISH   2 - ANY OTHER LANGUAGE
##                     1               94380                     7912
##        3 - DON'T KNOW          4 - REFUSED
##                    51                   9

table(d[, BMICLASS])

##
##          -2 - MISSING         1-UNDERWEIGHT
```

```
##              19963                      5921
##        2-NORMAL WEIGHT   3-AT RISK OF OVERWEIGHT
##              45650                     12207
##        4-OVERWEIGHT
##              18612

table(d[, S1Q01])
##
## -4 - PARTIAL INTERVIEW -3 - NOT IN UNIVERSE -2 - MISSING
##                      0                    0            1
##   -1 - LEGITIMATE SKIP           1 - MALE   2 - FEMALE
##                      0              52554        49719
##        6 - DON'T KNOW          7 - REFUSED
##                     14                   65
```

Recoding Data

In addition to programmatically finding cases with values to recode to missing (e.g., 7 - REFUSED in S1Q01 and 97 - REFUSED in EDUCATIO), we also want to drop those levels from the factor after converting them to missing. We can solve this by using regular expressions to search for the character strings we know. For example, we know REFUSED always ends in that regardless of whether it starts with 7 - or 97 - or anything else. We can search using regular expressions via the grep() function, which returns the value or the numeric position of matches, or the grepl() function, which returns a logical (hence the l) vector of whether each element matched or not.

Note   Regular expressions are a powerful tool for finding and matching character string patterns. In R, grep() and grepl() return the positions or a logical vector of which vector elements match the expected pattern. Combined with replacement, regular expressions help recode data.

We show these two functions in action with a simple example. The first formal, pattern, is the regular expression used for matching, while x is the data to search through. Notice the difference between logical vector and numeric position. Either way, the second and fourth locations of x contain our matching letters of abc:

```
grep(
  pattern = "abc",
  x = c("a", "abc", "def", "abc", "d"))
```

```
## [1] 2 4
```

```
grepl(
  pattern = "abc",
  x = c("a", "abc", "def", "abc", "d"))
```

```
## [1] FALSE TRUE FALSE TRUE FALSE
```

So far, we have a basic regular expression. We could use this for our data, searching for REFUSED instead of abc, although it is good to be as accurate as possible. For example, what if one variable has 1 - REFUSED TREATMENT, 2 - DID NOT REFUSED TREATMENT, 3 - REFUSED (ignoring the wonky example grammar), where 1 and 2 are valid responses and 3 indicates a refusal to answer the question. A search for REFUSED would match all of those. We ultimately want to use grepl(), as logical values are useful indexes to set specific cases to NA. To ensure we are grabbing the right values while checking that our regular expression is accurate, we use grep() with value = TRUE, which returns the matching strings. This makes it easy to see what we are matching:

```
grep(
  pattern = "REFUSED",
  x = c(
    "1 - REFUSED TREATMENT",
    "2 - DID NOT REFUSED TREATMENT",
    "3 - REFUSED"),
  value = TRUE)
```

```
## [1] "1 - REFUSED TREATMENT"         "2 - DID NOT REFUSED TREATMENT"
## [3] "3 - REFUSED"
```

To make it more specific, we could add the hyphen (-), which, although more precise, still gives us a false positive:

```
grep(
  pattern = "- REFUSED",
  x = c(
    "1 - REFUSED TREATMENT",
```

```
    "2 - DID NOT REFUSED TREATMENT",
    "3 - REFUSED"),
  value = TRUE)
```

```
## [1] "1 - REFUSED TREATMENT" "3 - REFUSED"
```

To go further, we specify that - REFUSED must be the last part of the string. That is, nothing can come after - REFUSED. This is done by using the $ in regular expressions to signify the *end* of the pattern string:

```
grep( pattern = "- REFUSED$",
  x = c(
    "1 - REFUSED TREATMENT",
    "2 - DID NOT REFUSED TREATMENT",
    "3 - REFUSED"),
  value = TRUE)
```

```
## [1] "3 - REFUSED"
```

Now, your data may be different, so what we need is a variety of ways to refine just what type of matching we need to have. Regular expressions allow us to specify what we expect to find. The following pattern, although short, contains much information. The [0-9] signifies the digits 0–9, and the + means the previous expression should occur at least one or more times. This code allows for both 1 and 97 yet not a blank and must be a number. After one or more numbers, the regular expression indicates we expect to find - REFUSED and then the end of the string, which we specify using the dollar sign. We extend our possible values to match for our x value and specify our pattern precisely, with the results shown here:

```
grep( pattern = "[0-9]+ - REFUSED$",
  x = c(
    "1 - TREATMENT REFUSED TREATMENT",
    "2 - DID NOT REFUSED TREATMENT",
    "3 - JACK - REFUSED",
    "4 - REFUSED",
    "97 - REFUSED",
    "- REFUSED"),
  value = TRUE)
```

```
## [1] "4 - REFUSED" "97 - REFUSED"
```

As a final step, we know that if we want genuine REFUSED in our real data set, the string starts with a number as well. Because it may be a negative number, we expand our previous expression to search for a negative operator zero or more times at the start of the string, followed by a number one or more times. We indicate the start of the string by using ^, and we indicate zero or more times by using *. By using this regular expression, we help protect against matching legitimate values in the data and errantly converting them to NA. We are quite specific in what we want, and anything that does not match our pattern is rejected:

```
grep( pattern = "^[-]*[0-9]+ - REFUSED$",
  x = c(
    "1 - TREATMENT REFUSED TREATMENT",
    "2 - DID NOT REFUSED TREATMENT",
    "3 - JACK - REFUSED",
    "4 - REFUSED",
    "97 - REFUSED",
    "-97 - REFUSED",
    "- REFUSED",
    "TRICK CASE 4 - REFUSED"),
  value = TRUE)

## [1] "4 - REFUSED"    "97 - REFUSED"    "-97 - REFUSED"
```

By default, regular expressions are also case sensitive. A full treatment of using regular expressions is beyond the scope of this book. We recommend *Mastering Regular Expressions* by Jeffrey Friedl (O'Reilly Media, 2006) for readers who are interested in comprehensive coverage of regular expressions. Regular expressions are quite useful for matching and working with string data because they allow us to encode very specific searches.

Going back to our data, we want to match more than just the word REFUSED. One option would be to create individual regular expressions for each and loop through, although this is cumbersome and inefficient. Instead, we can modify our regular expression to indicate different options in some positions. This is done by using a pipe, |, which functions as *or*, to separate options. In the following code, we keep the specifics about the start of the string but allow REFUSED or MISSING to be the ending text:

```
grep(
  pattern = "^[-]*[0-9]+ - REFUSED$|MISSING$",
  x = c(
    "1 - TREATMENT REFUSED TREATMENT",
    "2 - DID NOT REFUSED TREATMENT",
    "3 - JACK - REFUSED",
    "4 - REFUSED",
    "97 - REFUSED",
    "-97 - REFUSED",
    "-2 - MISSING",
    "- REFUSED",
    "TRICK CASE 4 - REFUSED",
    "4 - REFUSED HELP"),
  value = TRUE)
```

```
## [1] "4 - REFUSED"   "97 - REFUSED"   "-97 - REFUSED"   "-2 - MISSING"
```

Now we are ready to replace responses with NA for the patterns we want. As you see, the regular expression has become something rather complex. A word of caution is that many symbols have special meanings in regular expressions. Searching for special symbols requires special care. For instance, you have seen * is used to indicate zero or more times, not a literal asterisk. To search for a literal character, it must be escaped by using a backslash (\). We now build our pattern that ought to identify all missing values we would like to see in our data set on child health. We pull our pattern out as a separate piece for readability and run one last test on our made-up x data.

It is important to note the regular expression assigned to p has a line break in order to fit in the book, but it should be written as a single line with no space or line break before the NOT IN UNIVERSE$ portion:

```
p <- "^[-]*[0-9]+ - REFUSED$|MISSING$|DON'T KNOW$|LEGITIMATE SKIP$|PARTIAL
INTERVIEW$|NOT IN UNIVERSE$"
```

```
grep( pattern = p,
  x = c(
    "1 - TREATMENT REFUSED TREATMENT",
    "2 - DID NOT REFUSED TREATMENT",
    "3 - JACK - REFUSED",
```

```
    "4 - REFUSED",
    "97 - REFUSED",
    "-97 - REFUSED",
    "-2 - MISSING",
    "96 - DON'T KNOW",
    "-4 - LEGITIMATE SKIP",
    "-3 - PARTIAL INTERVIEW",
    "-2 - NOT IN UNIVERSE",
    "- REFUSED",
    "TRICK CASE 4 - REFUSED",
    "4 - PARTIAL INTERVIEW OF DOCTOR"),
  value = TRUE)
```

```
## [1] "4 - REFUSED"          "97 - REFUSED"
## [3] "-97 - REFUSED"        "-2 - MISSING"
## [5] "96 - DON'T KNOW"      "-4 - LEGITIMATE SKIP"
## [7] "-3 - PARTIAL INTERVIEW" "-2 - NOT IN UNIVERSE"
```

Now that we know how to find the cases we want to set to missing, it is time to act! Rather than type each variable, we loop through them in data.table, setting matching cases to NA. If it is a factor, we use droplevels() to remove unused factor levels and otherwise return them as is. To replace each variable, we use the := operator introduced in the previous chapter in a new way. Previously, you saw how to create a variable in a data table as dat[, NewVar := value]. Now we are going to replace multiple variables at once. To do this, we pass a vector of variable names on the left of the assignment operator, :=, and a list of the values on the right-hand side. Because the variables already exist in the data set, we are overwriting them rather than creating new ones. The variable names are stored in the vector, v. We wrap v in parentheses so that data.table evaluates v as an R object name containing a vector of variable names. If we did not wrap v in parentheses, data.table would try to assign the results to a variable named v.

The next challenge is selecting the variables. We normally type unquoted variable names in a data table, but our variable names are stored in the vector v. To accomplish this, we leverage an internal object in data.table, .SD, which is essentially a list version of the data table. However, we do not want to loop through every single variable in the data table. Although .SD defaults to being all variables (or columns) in the data table, we can make .SD contain only a subset of the variables in the data table by using the

.SDcols argument. .SDcols is a formal argument that takes a character vector and uses that vector to control the variables included in the internal .SD object in a data table. Thus, we specify the variables to include in .SD by writing .SDcols = v. This operation does not return any easy-to-read values, so instead we return to EDUCATIO and S1Q01:

```
v <- c("EDUCATIO", "PLANGUAG", "BMICLASS", "S1Q01", "S2Q01")
d[, (v) := lapply(.SD, function(x) {
  x[grepl(pattern = p, x)] <- NA
  if (is.factor(x)) droplevels(x) else x
}), .SDcols = v]

table(d[, EDUCATIO])
```

```
##
##           1 - LESS THAN HIGH SCHOOL 2 - 12 YEARS, HIGH SCHOOL GRADUATE
##                               4661                               21238
##           3 - MORE THAN HIGH SCHOOL
##                              76022
```

```
table(d[, S1Q01])
```

```
##
##   1 - MALE 2 - FEMALE
##      52554      49719
```

Note there is some inefficiency because the data is copied each time for the function run in the data table. It would be more efficient to select the relevant rows in data.table and then set those as missing. The difficulty here is our results are factors, and we want to drop the excess levels or coerce them to characters. We did not coerce them to factors in the first place; the data came that way from the Stata data file. This is simply part of dealing with the data we get. If we had all character data, we might have used a slightly more efficient process. If we wished, we might also have converted all factors to the character class at the beginning.

Note   The gsub() function combines matching with regular expressions and replaces the portions of a string that match the regular expression with new text in this form: gsub(pattern = "regex", replacement = "new values", x = "original data").

Another recoding task we may want is to drop the numbers so only the labels remain. This can be accomplished using the gsub() function, which performs regular expression matching within a string and then replaces matches with specified values (which can be a zero-length string). We first look at the following simple example:

```
gsub(
  pattern = "abc",
  replacement = "",
  x = c("a", "abcd", "123abc456"))
```

```
## [1] "a"        "d"        "123456"
```

Because it just uses regular expressions to match and remove numbers, we can reuse our pattern from before to test this out:

```
p.remove <- "^[-]*[0-9]+ - "
gsub(
  pattern = p.remove,
  replacement = "",
  x = c(
    "1 - TREATMENT REFUSED TREATMENT",
    "2 - DID NOT REFUSED TREATMENT",
    "3 - JACK - REFUSED",
    "4 - REFUSED",
    "97 - REFUSED",
    "-97 - REFUSED",
    "-2 - MISSING",
    "96 - DON'T KNOW",
    "-4 - LEGITIMATE SKIP",
    "-3 - PARTIAL INTERVIEW",
    "-2 - NOT IN UNIVERSE",
    "- REFUSED",
    "TRICK CASE 4 - REFUSED",
    "4 - PARTIAL INTERVIEW OF DOCTOR"))
```

```
##  [1] "TREATMENT REFUSED TREATMENT" "DID NOT REFUSED TREATMENT"
##  [3] "JACK - REFUSED"              "REFUSED"
##  [5] "REFUSED"                     "REFUSED"
```

```
##  [7] "MISSING"                    "DON'T KNOW"
##  [9] "LEGITIMATE SKIP"            "PARTIAL INTERVIEW"
## [11] "NOT IN UNIVERSE"            "- REFUSED"
## [13] "TRICK CASE 4 - REFUSED"     "PARTIAL INTERVIEW OF DOCTOR"
```

For efficiency, we can combine this with our previous code to set some cases to missing. Doing it in one step avoids repetitive processing. First, though, we read the data back in so we have the original raw data:

```
d <- read.dta("data/ch08/ICPSR_04691/DS0001/04691-0001-Data.dta")
d <- as.data.table(d)
setkey(d, IDNUMR)
```

Because gsub() coerces the input to characters regardless of whether we pass factors or characters to it, we will get characters out. So to know whether the input was a factor or not, we create an object, f, which is a logical indicating whether x started off as a factor. Then, rather than drop levels, we just convert them to factors if appropriate. Other than that change, we meld our earlier code with our removal code:

```
d[, (v) := lapply(.SD, function(x) {
  f <- is.factor(x)
  x[grepl(pattern = p, x)] <- NA
  x <- gsub(pattern = p.remove, replacement = "", x)
  if (f) factor(x) else x
}), .SDcols = v]

table(d[, EDUCATIO])

##
## 12 YEARS, HIGH SCHOOL GRADUATE           LESS THAN HIGH SCHOOL
##                           21238                            4661
##          MORE THAN HIGH SCHOOL
##                           76022

table(d[, S1Q01])

##
## FEMALE    MALE
##  49719   52554
table(d[, S2Q01])
```

```
##
## EXCELLENT      FAIR     GOOD      POOR VERY GOOD
##     65252      2189    10680       297     23903
```

Recoding Numeric Values

Because they have a more systematic structure, recoding numeric values is easier than recoding string data. To indicate different types of "missingness" in numeric data, people often use out-of-bounds values, such as a negative number or very high numbers (e.g., 999). Coding different types of missing values makes sense from a data perspective, as the codes provide additional information (e.g., skipped a question, not an applicable question). However, from a practical use perspective, we typically want to ignore all missing data (e.g., when calculating the mean, having data using -2 as a child's age when the parent refused to report it or -3 if a parent forgot will hurt our output). We can find values that fall between a range by using the %between% operator in the data.table package, and we can locate the complement by using negation !.

For example, for height, the documentation indicates values zero or below, and above 90, are used to code various types of missing. Similarly, for weight, values zero or below, as well as above 900, are used to code different types of missing values. To tell R these values indicate missing, we need to recode them. We see from our table() data there are indeed some missing values, and we can see exactly what values are used to code them (-2, 96, 96, 996, 997):

```
table(d[!S2Q02R %between% c(0, 90), S2Q02R])

##
##   -2   96   97
##    4 8096  131

table(d[!S2Q03R %between% c(0, 900), S2Q03R])

##
##   -2  996 997
##    4 2375  80
```

Again we could operate on these variables individually. However, doing so becomes cumbersome for many variables. The pattern often is that if valid values are less than 9, then 9 is missing. If values go into double digits, then ¿90 is missing, and so on. We recode programmatically by examining the maximum value and setting the bounds accordingly. We see what the variable was like originally with the following code and in Figure 8-1.

```
par(mfrow = c(1,2))
hist(d$S2Q02R)
hist(d$S2Q03R)
```

We focus on three columns needing numeric recoding. For each variable, we loop through calling the column name in our list. Designating m as our maximum, we start off with j and i set equal to 1. As long as j is equal to 1 and i is less than or equal to the number of values to try (e.g., 9, 99, 999), we keep doing the next calculation. If the maximum (ignoring missing) of the variable, k, is less than the maximum of the ith m value, then set j to 0 (which breaks the loop) and replace any values outside the range of 0 and the ith m. If the ith m is greater than 90 or minus some minuscule number, set the variable to missing (NA_integer_). In essence, we are counting through our list by columns, and we check whether the maximum value in our entire column is one of our maximum m values. Once it is there, we make sure that, effectively, values from 9 to 10 or 90 to 100 or 900 to 1000 are set to missing. Note we use NA_integer_ rather than NA because we are replacing a subset of values in an existing variable. data.table expects the class of data being used to replace values within an existing variable matches the class of that variable. We could use NA earlier because we overwrote the entire variable rather than replacing select values from within the variable:

```
v2 <- c("S2Q02R", "S2Q03R", "AGEYR_CH")
m <- sort(c(9, 99, 999))

for (k in v2) {
  j <- i <- 1
  while(j == 1 & i <= length(m)) {
    if (max(d[[k]], na.rm = TRUE) < m[i]) {
      j <- 0
      d[!(get(k) %between% c(0, ifelse(m[i] > 90,
        m[i] - 9, m[i] - 1e-9))), (k) := NA_integer_]
```

```
  } else {
    i <- i + 1
  }
  }
}
```

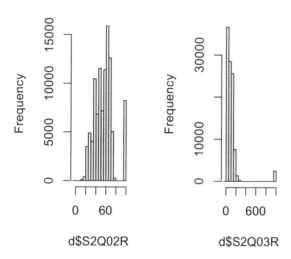

Figure 8-1. *Outliers that are near >900 and >90*

We show the result of this after recoding by using the same histogram function as before and in Figure 8-2 and Figure 8-3. Naturally the x-axis is much tighter after recoding:

hist(d$S2Q02R)

hist(d$S2Q03R)

With this, we conclude our section about data hygiene. Next, we show how to create new variables to simplify the analysis process. In a data science perspective, variable creation is sometimes called feature creation. In many domains of social and behavioral research, questionnaires or surveys are commonly employed. In education, tests are administered. Regardless of whether the data is from tests, questionnaires, or surveys, a common task is to aggregate responses to individual items or variables to create a scale score, the overall test score, or some other aggregate index.

8.2 Creating New Variables

A questionnaire asks whether a doctor or health professional ever told the respondent that the focal child had any of nine possible conditions (e.g., asthma, ADD or ADHD, depression or anxiety problems, diabetes, developmental delay, or physical impairment). Composite variable creation is the goal, to capture a number of issues. Some problems apply only to older children, and some respondents may not know the answer or may have refused to answer. Thus, we might take the average number of yes responses to calculate the proportion of yes responses out of all valid responses per respondent.

We first create a list of variables that pulls these nine columns' worth of data. From there, to get our actual variables, we unlist() them from our data table to see all possible responses that may require cleanup to reduce to a yes/no scenario. Notice in the following code there are several cases of missing data, legitimate skips, don't know, and refused:

```
v.health <- paste0("S2Q", c(19, 20, 21, 22, 23, 24, 26, 35, 37))
v.health
```

```
## [1] "S2Q19" "S2Q20" "S2Q21" "S2Q22" "S2Q23" "S2Q24" "S2Q26" "S2Q35"
## [9] "S2Q37"
```

```
table(unlist(d[, v.health, with = FALSE]))
```

```
##
## -4 - PARTIAL INTERVIEW    -3 - NOT IN UNIVERSE        -2 - MISSING
##                        0                       0                 120
##    -1 - LEGITIMATE SKIP              0 - NO             1 - YES
##                    48388                  833454               37720
##         6 - DON'T KNOW           7 - REFUSED
##                     1362                     133
```

Histogram of d$S2Q02R

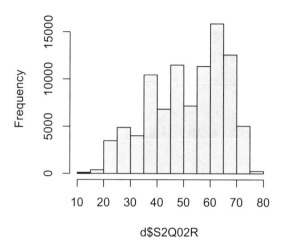

Figure 8-2. *The outliers are gone because histograms ignore missing values*

Histogram of d$S2Q03R

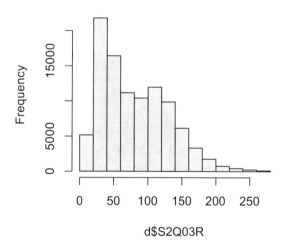

Figure 8-3. *The outliers are gone because histograms ignore missing values*

We already know how to clean these sorts of variables by using regular expressions and grep(). After cleanup, we check all the responses again, including NAs, in our table with the argument useNA = "ifany". It is important to note that although the regular expression has a line break to fit in this book, it should be written as a single line with no space or line break before the NOT IN UNIVERSE$ portion:

```
p <- "^[-]*[0-9]+ - REFUSED$|MISSING$|DON'T KNOW$|LEGITIMATE SKIP$|PARTIAL
INTERVIEW$|NOT IN UNIVERSE$"

d[, (v.health) := lapply(.SD, function(x) {
  x[grepl(pattern = p, x)] <- NA
  if (is.factor(x)) droplevels(x) else x
}), .SDcols = v.health]

table(unlist(d[, v.health, with = FALSE]), useNA = "ifany")

##
##  0 - NO 1 - YES   <NA>
## 833454   37720  50003
```

Now we can create a new variable that is the average of yes responses. To calculate the average by row, we use the rowMeans() function. We leverage the fact that logical values are stored as FALSE = 0 and TRUE = 1. Thus, we test whether the character data is equal to 1 - YES, which returns a boolean, and then take the row means of those logical values. We ignore (drop) missing values by setting na.rm = TRUE:

```
d[, HealthConditions := rowMeans(d[, v.health, with = FALSE] == "1 - YES",
na.rm = TRUE)]

table(round(d$HealthConditions, 2),
      useNA = 'ifany')

##
##      0  0.11 0.12 0.14 0.17  0.2 0.22 0.25 0.29 0.33 0.38
## 77426 16116  239   20    5  676 4386  156   11 1933   75
##    0.4 0.43 0.44  0.5 0.56 0.57  0.6 0.62 0.67 0.71 0.75
##     81   12  768   51  264    3    3   13   80    1    5
##   0.78 0.88 0.89  NaN
##     19    1    3    6
```

We could follow a similar process to get the row counts. Looking at the counts, something has happened, though. There are no missing values! This is because with rowSums(), when na.rm = TRUE and a row has no valid data, the sum is zero, not a missing value. To correct this, we need to manually set to missing any cases/rows where all are missing:

```
d[, NHealthConditions := rowSums(d[, v.health, with = FALSE] == "1 - YES",
na.rm = TRUE)]
```

```
table(d$NHealthConditions, useNA = 'ifany')
```

```
##
##     0     1     2     3     4     5     6     7     8
## 77432 17068  4630  2018   820   278    84    20     3
```

We count the number of missing responses per respondent, select rows where the number missing equals the total, and set those to missing. Again note that when replacing a subset of values from a variable in data.table, match the class of missing to the class of the variable (e.g., NA_integer_ for integer, NA_real_ for numeric, NA_character_ for character, NA for logical). That gives us a better count:

```
d[, NMissHealthConditions := rowSums(is.na(d[, v.health, with = FALSE]))]
```

```
d[NMissHealthConditions == length(v.health), NHealthConditions := NA_integer_]
```

```
table(d$NHealthConditions, useNA = 'ifany')
```

```
##
##     0     1     2     3     4     5     6     7     8   <NA>
## 77426 17068  4630  2018   820   278    84    20     3      6
```

Using rowMeans() or rowSums() to create new variables is a relatively elegant solution. However, it is not a very data.table approach. The calculation occurs outside the original data.table; we call the object, d, a second time and subset the columns for use. When calling the data object a second time, we lose access to many features of data.table, such as performing operations for only certain subsets of the data or performing an operation by some grouping factor.

If ignoring missing values is not an issue, we can easily work directly with data.table by using the Reduce() function. Reduce() takes a function (typically a binary operator) as its first argument and a vector or list of arguments. We show examples here of a vector and of a list with addition, division, and powers:

```
Reduce('+', c(1, 2, 3))
```

```
## [1] 6
```

```
Reduce('+', list(1:3, 4:6, 7:9))
```

```
## [1] 12 15 18
```

```
Reduce('*', list(1:3, 4:6, 7:9))
```

```
## [1]  28  80 162
```

```
Reduce('^', list(1:3, 4:6, 3:1))
```

```
## [1]    1  1024  729
```

```
Reduce('-', list(1:3, 4:6, 7:9))
```

```
## [1] -10 -11 -12
```

```
Reduce('/', list(1:3, 4:6, 7:9))
```

```
## [1] 0.036 0.050 0.056
```

This can then easily be applied to variables in a data.table. We cannot directly add the health variables, because they are factor class data. However, we could write a function to deal with it:

```
fplus <- function(e1, e2) {
  if (is.factor(e1)) {
    e1 <- as.numeric(e1) - 1
  }
  if (is.factor(e2)) {
    e2 <- as.numeric(e2) - 1
  }
e1 + e2
}
```

We select the columns by using .SDcols, and then .SD will be a list of variables we can reduce and store results in a new variable:

```
d[, NHealthConditions2 := Reduce(fplus, .SD), .SDcols = v.health]
```

```
table(d$NHealthConditions2)
```

```
##
##      0     1     2     3     4     5     6     7     8
## 65459 16116  4386  1927   768   264    79    19     3
```

8.3 Fuzzy Matching

Approximate string matching, or *fuzzy* matching, is a technique whereby observations link to a reference list. For example, you may have a list of registered users, and then individuals write their names on an attendance sheet. Alternately, you may be working with filenames meant to match certain pieces of information.

We start with two lists; the reference names are our registered user list. These are the names we believe in and want to match to the observed names that were written in quickly on our hypothetical attendance sheets. Users may have attended more than one event and thus show up multiple times or may not show up at all. Also note some of these words have double spaces between them, which are required to match our later output:

```
reference.Names <- c("This Test", "Test Thas",
  "Jane Mary", "Jack Dun-Dee")
observed.Names <- c("this test", "test this", "test that",
  "JaNe Mary.", "Mary Sou",  "Jack Dee", "Jane Jack")
```

The challenge is to find out the number of events registered users attended (i.e., signed in to the sheet). A first pass can be done by using the `stringdistmatrix()` function from the `stringdist` package. We use the Damerau-Levenshtein distance method, which essentially counts the number of characters that have to change to go from one string to another. From this, we can see for each observed name the minimum number of characters that must be changed to match one of our reference names. For this test, it takes two character changes to match the first reference name, eight for the second reference name, and so on. Note that by default, `stringdistmatrix()` is case sensitive, so switching cases counts as one change. Notice this matrix gives us a grid of those matches, where the matrix elements are the distance:

```
stringdistmatrix(reference.Names,
                 observed.Names,
                 method = "dl")
```

```
##       [,1] [,2] [,3] [,4] [,5] [,6] [,7]
## [1,]    2    8    7   10    8    7    8
## [2,]    8    3    3    9    8    8    8
## [3,]    8    8    8    3    7    6    3
## [4,]   11   11   11    9   10    4    9
```

If we want only matches, we can use approximate matching. It takes similar arguments, but also the maximum distance allowed before calling nothing a match. Notice it returns positions in the reference name vector:

```
amatch(observed.Names,
       reference.Names,
       method = "dl", maxDist = 4)
```

```
## [1] 1   2  2  3  NA  4  3
```

We can expand a bit and make it non-case sensitive by converting to lowercase via `tolower()` or to uppercase via `toupper()`. Additionally, we can remove some text items that should not be part of the match, such as various punctuation marks. Both these methods have a chance to give us lower distances overall. Although in our case there are no periods to remove, other text hygiene could be performed using the `gsub` methods learned earlier.

```
stringdistmatrix(
  tolower(reference.Names),
  tolower(observed.Names),
  method = "dl")
```

```
##        [,1] [,2] [,3] [,4] [,5] [,6] [,7]
## [1,]    0    6    5   10    8    7    8
## [2,]    6    1    1    9    8    8    8
## [3,]    8    8    8    2    7    6    3
## [4,]   11   11   11    9   10    4    9
```

```
stringdistmatrix(
  tolower(gsub("\\.", "", reference.Names)),
  tolower(gsub("\\.", "", observed.Names)), method = "dl")
```

```
##        [,1] [,2] [,3] [,4] [,5] [,6] [,7]
## [1,]    0    6    5    9    8    7    8
## [2,]    6    1    1    8    8    8    8
## [3,]    8    8    8    1    7    6    3
## [4,]   11   11   11    9   10    4    9
```

We can split a character string by a particular character by using strsplit(). Any pattern or character can be used, but we use spaces. We use backslash s (\s) as a special character in regular expressions to mean any type of space. A second backslash is required in order to escape the first backslash. That is because the special character is not s, which would be escaped as \s, but rather the special character is the compound \s—thus, the escaped version is \\s. We demonstrate the result of strsplit() in the following code:

```
strsplit(
  x = reference.Names[1],
  split = "\\s")[[1]]
```

```
## [1] "This" "Test"
```

```
strsplit(
  x = observed.Names[2],
  split = "\\s")[[1]]
```

```
## [1] "test" "this"
```

Now we can use stringdistmatrix() again. The result shows each sub-chunk of the second observed name is one character manipulation away from one of the sub-chunks from the first reference name:

```
stringdistmatrix(
  strsplit(
    x = reference.Names[1],
    split = "\\s")[[1]],
  strsplit(
    x = observed.Names[2],
    split = "\\s")[[1]],
  method = "dl")
```

```
##      [,1] [,2]
## [1,]    4    1
## [2,]    1    4
```

If we ignore case, we get perfect matches. The second chunk of observed name 2 matches the first chunk of reference name 1, and the first chunk of observed name 2 perfectly matches the second chunk of the reference name 1:

```
stringdistmatrix(
  strsplit(
    x = tolower(reference.Names[1]),
    split = "\\s")[[1]],
  strsplit(
    x = tolower(observed.Names[2]),
    split = "\\s")[[1]],
  method = "dl")
```

```
##        [,1] [,2]
## [1,]    3    0
## [2,]    0    3
```

To return the sum of the best matches, we can pick the minimum and sum. The results show us ignoring case and ignoring ordering makes the second observed name a perfect match for the first reference name. We would never have gotten this result if we tested it as an overall string:

```
sum(apply(stringdistmatrix(
  strsplit(
    x = tolower(reference.Names[1]),
    split = "\\s")[[1]],
  strsplit(
    x = tolower(observed.Names[2]),
    split = "\\s")[[1]],
  method = "dl"), 1, min))
```

```
## [1] 0
```

Combining everything discussed yields a list of techniques to use for string matching. We find values to ignore (e.g., punctuation), split strings (e.g., on spaces), and ignore case. We write a function to help with matching names. To this, we also add an optional argument, fuzz, to control whether to include close matches within a certain degree of tolerance. We also count exact matches. This code is not optimized for speed or efficiency, but is an example combining many aspects of what you've learned. It also applies this knowledge to a set of names, where each written name (index argument) compares against a vector of possible candidates (the pool). We comment the function inline to explain the pieces, rather than write a wall of text at the beginning:

```r
matchIt <- function(index,
                     pool,
                     ignore = c("\\.", "-"),
                     split = FALSE,
                     ignore.case = TRUE,
                     fuzz = .05,
                     method = "dl") {
  # for those things we want to ignore, drop them
  # e.g., remove spaces, periods, dashes
  rawpool <- pool

  for (v in ignore) {
    index <- gsub(v, "", index)
    pool <- gsub(v, "", pool)
  }
  # if ignore case, convert to lower case
  if (ignore.case) {
    index <- tolower(index)
    pool <- tolower(pool)
  }

if (!identical(split, FALSE)) {
  index2 <- strsplit(index, split = split)[[1]]
  index2 <- index2[nzchar(index2)]
  pool2 <- strsplit(pool, split = split)
  pool2 <- lapply(pool2, function(x)
    x[nzchar(x)])

  # calculate distances defaults to the Damerau-Levenshtein distance
  distances <- sapply(pool2, function(x) {
    sum(apply(
      stringdistmatrix(index2, x, method = method),
      1,
      min,
      na.rm = TRUE
    ))
  })
```

```
} else {
  # calculate distances defaults to the Damerau-Levenshtein distance
  distances <- stringdist(index, pool, method = method)
  }

  # some methods result in Infinity answers, set these missing
  distances[!is.finite(distances)] <- NA

  # get best and worst
  best <- min(distances, na.rm = TRUE)
  worst <- max(distances, na.rm = TRUE)

  # if fuzz, grab all distances that are within fuzz percent of
  # the difference between best and worst, tries to capture
  # potentially very close matches to the best
  if (fuzz) {
    usedex <- which(distances <= (best + ((worst - best) * fuzz)))
  } else {
    usedex <- which(distances == best)
  }

  # define a distance below which its considered a
  # perfect or exact match
  perfect <- distances[usedex] < .01
  out <- rawpool[usedex]

  # count the number of perfect matches
  count <- sum(perfect)

  if (any(perfect)) {
    # if there are perfect matches, just use one
    match <- out[perfect][1]

    # return a data table of the perfect match
    # and number of perfect matches (probably 1)
    data.table(Match = match,
               N = count,
```

```
                Written = NA_character_)
  } else {
    # if no perfect matches,
    # return a list of close matches, comma separated
    # also return what exactly was written
    data.table(
      Match = paste(out, collapse = ", "),
      N = count,
      Written = index
    )
  }
}
```

Now we loop through each observed name and try to match it to reference names within 5 percent of the best match (.05). Since the output is always the same, we can combine it row-wise:

```
output <- lapply(observed.Names, function(n) {
  matchIt(
    index = n,
    pool = reference.Names,
    ignore = c("\\.", "-"),
    split = "\\s",
    fuzz = .05)
})

output <- do.call(rbind, output)

output

##                            Match N   Written
## 1:                     This Test 1     <NA>
## 2:                     This Test 1     <NA>
## 3:                     Test Thas 0 test that
## 4:                     Jane Mary 1     <NA>
## 5:                     Jane Mary 0  mary sou
## 6:                 Jack Dun-Dee 0  jack dee
## 7: Jane Mary, Jack Dun-Dee 0 jane jack
```

We see This Test shows up in two rows, so we aggregate up and print the final output. There are two perfect matches for This Test. There is no perfect match for Test Thas, although there is a close one listed under the Uncertain column. Finally, jack dee is a close but uncertain match for Jack Dun-Dee, and jane jack is an uncertain match for Jane Mary or Jack Dun-Dee:

```
finaloutput <- output[, .(
   N = sum(N),
   Uncertain = paste(na.omit(Written), collapse = ", ")),
   by = Match]

finaloutput

##                        Match N Uncertain
## 1:                  This Test 2
## 2:                  Test Thas 0 test that
## 3:                  Jane Mary 1  mary sou
## 4:              Jack Dun-Dee 0  jack dee
## 5: Jane Mary, Jack Dun-Dee 0 jane jack
```

Although it is challenging to cover every example and use case, these basic tools should be enough to cover many types of data management tasks when used in combination.

In particular, variations on fuzzy matching are used by both authors for various scenarios involving student peer-ranking (on classroom activities) and sign-in sheet matching (for recruitment and professional development rosters).

8.4 Summary

In this chapter, you saw how to locate text based on either exact regular expressions or fuzzy matching. Additionally, you learned how to substitute new string pieces for old. Finally, we discussed ways of creating new variables in data.table. Table 8-1 summarizes the key functions presented in this chapter.

Table 8-1. *Listing of key functions described in this chapter and summary of what they do*

Function	What It Does
`read.dta()`	Reads Stata files.
`str()`	Displays R object structure.
`grep()`	Pattern matching that returns a vector showing which elements match.
`grepl()`	Pattern matching that returns a vector showing boolean for all values in x.
`gsub()`	Matches a pattern and makes a substitution by returning the original vector after making changes.
`rowMeans()`, `rowSums()`	Calculates the average or sum of values for each row of a data table or matrix.
`Reduce()`	Takes a binary function as its first argument and applies that to its remaining argument
`.SD`	Reserved name within data.table that can be used to refer to all the variables within the data table as a list.
`.SDcols`	Argument to data.table that allows specification of the variables to be included in .SD.
`strsplit()`	Splits a string based on a particular splitting character.
`stringdistmatrix()`	Creates a matrix that has distances between observations and expected values.
`amatch()`	Provides a vector that shows which index in the expected values most closely matches the observed values.
`tolower()`	Coerces all characters in the given string to lowercase.

CHAPTER 9

Other Data Management Packages

Comparing data frames and data tables leads to an interesting question. What if there were more types of data? Particularly, what if there were different ways to store data that are all, at their heart, tables of some sort? In addition to data frames or tables, there are many ways to store data, many of which are just tables. The idea behind dplyr [29] is that regardless of what the data back end might be, our experience should be the same. To allow this, dplyr implements generic functions for common data management tasks. For each of these generic functions, specific methods are written that translate the generic operation into whatever code or language is required for a specific back end. Using a layer of abstraction ensures that users get a consistent experience, regardless of the specific data format, or back end, being used. It also makes dplyr extensible, in that support for a new format can be added by simply writing additional functions or methods. The user experience need not change.

Somewhat like data.table, the dplyr has its own enhancement of data frames called tidy tables or tibbles [12]. Combined with a set of data manipulation functions in the tidyr [30] package, these work together to allow for thoughtful manipulation and organization of data. Since the first edition of this book, this family of packages has come to be known as the tidy universe or *tidyverse* [26].

Part of this tidy philosophy of data science is naming functions after descriptive verbs denoting a specific action. That is, rather than have a few functions adaptable to many tasks, tidy has unique functions for the most common tasks. This makes code more human readable, even to those not familiar with R. While data.table's relentless focus on efficiency continues to make it the clear winner for speed, there is no denying some value in more intuitive code. Both approaches have strengths and challenges.

M. Wiley and J. F. Wiley, *Advanced R 4 Data Programming and the Cloud*,
https://doi.org/10.1007/978-1-4842-5973-3_9

Your authors are most familiar with data.table–so we may be forgiven for having a clear favorite. However, there are cases where tibbles either make more sense or are more approachable for a wider audience. Thus, it would not make sense to have a text about advanced uses of R without discussing the tidy approach.

As before, we set some options to reduce the number of rows printed. Rather than call dplyr, tibble, and tidyr separately, we call them in a single wrapper package named tidyverse:

```
library(tidyverse)
library(extraoperators)
library(stringdist)
options(stringsAsFactors = FALSE,
        tibble.print_max = 20, ## if over 20 rows
        tibble.print_min = 5, ## print first 5
        digits = 2) ## reduce digits printed by R
```

For data.table, we used sapply() to get a vector of classes for each column or variable. Although we could do the same with tibbles, tibbles automatically provide information about the types or classes of data in each column as well as total number of columns and rows (e.g., <dbl> indicating a *double* or numeric value, with <chr> indicating *character* class data). Despite the differences, there remain the common goals of working with data: sorting, subsetting, ordering, computing, and reshaping:

```
iris$Species <- as.character(iris$Species)
diris <- as_tibble(iris)
diris
```

```
## # A tibble: 150 x 5
##    Sepal.Length Sepal.Width Petal.Length Petal.Width Species
##           <dbl>       <dbl>        <dbl>       <dbl> <chr>
## 1           5.1         3.5          1.4         0.2 setosa
## 2           4.9         3            1.4         0.2 setosa
## 3           4.7         3.2          1.3         0.2 setosa
## 4           4.6         3.1          1.5         0.2 setosa
## 5           5           3.6          1.4         0.2 setosa
## # ... with 145 more rows
```

9.1 Sorting

A helpful way to think of tidy is it uses verbs to describe actions on the data. In the tidyverse, data are sorted by the arrange() function, which takes a data set as its first argument:

```
diris <- arrange(diris, Sepal.Length)
diris
```

```
## # A tibble: 150 x 5
##   Sepal.Length Sepal.Width Petal.Length Petal.Width Species
##          <dbl>       <dbl>        <dbl>       <dbl> <chr>
## 1          4.3           3          1.1         0.1 setosa
## 2          4.4         2.9          1.4         0.2 setosa
## 3          4.4           3          1.3         0.2 setosa
## 4          4.4         3.2          1.3         0.2 setosa
## 5          4.5         2.3          1.3         0.3 setosa
## # ... with 145 more rows
```

Tidyverse works via functions that most often take the convention of a verb meant to signal the action being performed on the data. In order to follow the usual English sentence structure of noun-verb, these functions are also built to use the *pipe operator*. The pipe operator is coded as %>% and removes the need to reference the data set inside each function. In particular, it pipes the data object (noun) on the left-hand side into the *first* formal of the function (verb) on the right-hand side.

Compare and contrast the following code with the prior code to see how the first formal now lives on the left-hand side:

```
diris %>% arrange(Sepal.Length)
```

```
## # A tibble: 150 x 5
##   Sepal.Length Sepal.Width Petal.Length Petal.Width Species
##          <dbl>       <dbl>        <dbl>       <dbl> <chr>
## 1          4.3           3          1.1         0.1 setosa
## 2          4.4         2.9          1.4         0.2 setosa
## 3          4.4           3          1.3         0.2 setosa
## 4          4.4         3.2          1.3         0.2 setosa
## 5          4.5         2.3          1.3         0.3 setosa
## # ... with 145 more rows
```

It is worth noting that, in the event a function does not reserve the *first* formal for *data*, using a . (period or full stop) in that position of the right-hand side function arguments will allow the pipe to work as needed.

By itself, the pipe operator does not modify our dataset, `diris`. As always, the assignment operator would be required to make a permanent change. The `arrange()` function takes additional arguments after the first argument to provide sorting by those other columns. The default is an increasing order, and this is optionally modified with a second function call of `desc()` to provide *descending* order:

```
diris <- diris %>% arrange(Sepal.Length, desc(Sepal.Width))
diris
```

```
## # A tibble: 150 x 5
##    Sepal.Length Sepal.Width Petal.Length Petal.Width Species
##           <dbl>       <dbl>        <dbl>       <dbl> <chr>
## 1           4.3         3            1.1         0.1 setosa
## 2           4.4         3.2          1.3         0.2 setosa
## 3           4.4         3            1.3         0.2 setosa
## 4           4.4         2.9          1.4         0.2 setosa
## 5           4.5         2.3          1.3         0.3 setosa
## # ... with 145 more rows
```

It is possible to see the first-row index when a duplicate occurs for either the entire data set or just a particular column, via `anyDuplicated()`. When calling this function on the entire data set, it looks for a match between all columns, which occurs for rows 78 and 79 in this data set. A duplicated variable in `Sepal.Length` occurs in the third row, as shown in the following code:

```
diris %>% anyDuplicated()
```

```
## [1] 79
```

```
anyDuplicated(diris$Sepal.Length)
```

```
## [1] 3
```

Whereas anyDuplicated() returns the row index of the first duplication, duplicated() simply tells us how many such duplications occur in our table. Looking at our code, we see for a full match in all columns, row 79 is the only time that occurs. On the other hand, duplicated sepal lengths may happen for the first time in row 3, but there are many more duplicates:

```
diris %>% duplicated() %>% table()
```

```
## .
## FALSE   TRUE
##   149      1
```

```
table(duplicated(diris$Sepal.Length))
```

```
##
## FALSE   TRUE
##    35    115
```

If we want to see table results with only distinct values, distinct() works much like arrange(). The first formal is for data, while the rest specify column name variables. If no column names are specified, distinct() operates on all columns. Recall from the duplicated() call that there are 149 unique rows:

```
diris %>% distinct()
```

```
## # A tibble: 149 x 5
##    Sepal.Length Sepal.Width Petal.Length Petal.Width Species
##           <dbl>       <dbl>        <dbl>       <dbl> <chr>
## 1           4.3         3            1.1         0.1 setosa
## 2           4.4         3.2          1.3         0.2 setosa
## 3           4.4         3            1.3         0.2 setosa
## 4           4.4         2.9          1.4         0.2 setosa
## 5           4.5         2.3          1.3         0.3 setosa
## # ... with 144 more rows
```

Similarly, recall there are only 35 unique rows if we consider only sepal length. Consequently, using the second formal of distinct() to hold the Sepal.Length column yields only 35 rows in our tibble. We show the first five of those rows here:

```
diris %>% distinct(Sepal.Length)
```

```
## # A tibble: 35 x 1
##    Sepal.Length
##           <dbl>
## 1          4.3
## 2          4.4
## 3          4.5
## 4          4.6
## 5          4.7
## # ... with 30 more rows
```

This section closes with one last call to `distinct()`, showing more than one column called in the second and third formal arguments. Here, a row is unique provided both `Sepal.Length` and `Sepal.Width` do not duplicate. Pay special attention to rows 2–4, which have the same `Sepal.Length`, yet those rows are not duplicated because their `Sepal.Widths` are different:

```
diris %>% distinct(Sepal.Length, Sepal.Width)
```

```
## # A tibble: 117 x 2
##    Sepal.Length Sepal.Width
##           <dbl>       <dbl>
## 1          4.3           3
## 2          4.4         3.2
## 3          4.4           3
## 4          4.4         2.9
## 5          4.5         2.3
## # ... with 112 more rows
```

9.2 Selecting and Subsetting

Selecting portions of the stored data in `tidyverse` is different from the `data.table` methodology. Remember, `tidyverse` works via function calls. These functions are designed to be independent of the underlying data structure. Nevertheless, we do want the same end results. Selecting rows by position is accomplished using `slice()`, which takes two formal arguments. The first is the data object, in our case `diris`, and the second allows us to select specific rows. Keep in mind we are using the *pipe* operator, %>%. Thus, our first formal is *outside* our function! The two lines of code select just the first five rows, one in traditional function notation and the other using the pipe operator. It may seem slightly cleaner yet silly to use pipe right now. The true strength of pipe becomes clearer as we start performing more complex operations:

```
slice(diris, 1:5)
```

```
## # A tibble: 5 x 5
##    Sepal.Length Sepal.Width Petal.Length Petal.Width Species
##           <dbl>       <dbl>        <dbl>       <dbl> <chr>
## 1          4.3           3          1.1         0.1 setosa
## 2          4.4         3.2          1.3         0.2 setosa
## 3          4.4           3          1.3         0.2 setosa
## 4          4.4         2.9          1.4         0.2 setosa
## 5          4.5         2.3          1.3         0.3 setosa
```

```
diris %>% slice(1:5)
```

```
## # A tibble: 5 x 5
##    Sepal.Length Sepal.Width Petal.Length Petal.Width Species
##           <dbl>       <dbl>        <dbl>       <dbl> <chr>
## 1          4.3           3          1.1         0.1 setosa
## 2          4.4         3.2          1.3         0.2 setosa
## 3          4.4           3          1.3         0.2 setosa
## 4          4.4         2.9          1.4         0.2 setosa
## 5          4.5         2.3          1.3         0.3 setosa
```

As seen before, - can be used for negation. Therefore, we could also drop all rows but the last five rows by dropping rows 1–145:

```
diris %>% slice(-(1:145))
```

```
## # A tibble: 5 x 5
##    Sepal.Length Sepal.Width Petal.Length Petal.Width Species
##           <dbl>       <dbl>        <dbl>       <dbl> <chr>
## 1          7.7         3.8          6.7         2.2 virginica
## 2          7.7           3          6.1         2.3 virginica
```

## 3	7.7	2.8	6.7	2	virginica
## 4	7.7	2.6	6.9	2.3	virginica
## 5	7.9	3.8	6.4	2	virginica

Rather than select rows directly, we may set logical conditions on them. Again, these types of arguments belong in the second formal of our function call. However, in this case, we use filter() rather than slice(). The filter() function uses logical indexing. It is easy enough to select all the rows in which Species either is or is not setosa by writing a logical test:

```
diris %>% filter(Species == "setosa")
```

```
## # A tibble: 50 x 5
##    Sepal.Length Sepal.Width Petal.Length Petal.Width Species
##           <dbl>       <dbl>        <dbl>       <dbl> <chr>
## 1          4.3           3          1.1         0.1 setosa
## 2          4.4         3.2          1.3         0.2 setosa
## 3          4.4           3          1.3         0.2 setosa
## 4          4.4         2.9          1.4         0.2 setosa
## 5          4.5         2.3          1.3         0.3 setosa
## # ... with 45 more rows
```

```
diris %>% filter(Species != "setosa")
```

```
## # A tibble: 100 x 5
##    Sepal.Length Sepal.Width Petal.Length Petal.Width Species
##           <dbl>       <dbl>        <dbl>       <dbl> <chr>
## 1          4.9         2.5          4.5         1.7 virginica
## 2          4.9         2.4          3.3           1 versicolor
## 3            5         2.3          3.3           1 versicolor
## 4            5           2          3.5           1 versicolor
## 5          5.1         2.5            3         1.1 versicolor
## # ... with 95 more rows
```

It is also possible to select rows based on numeric values with inequalities, equality, or non-equality in addition to logical indexes. Both numeric and character tests may be mixed and matched to extract precisely the rows desired:

```
iris %>% filter(Sepal.Length < 5 & Petal.Width > .2)
```

```
##    Sepal.Length Sepal.Width Petal.Length Petal.Width   Species
## 1           4.6         3.4          1.4         0.3    setosa
## 2           4.5         2.3          1.3         0.3    setosa
## 3           4.8         3.0          1.4         0.3    setosa
## 4           4.9         2.4          3.3         1.0 versicolor
## 5           4.9         2.5          4.5         1.7  virginica
```

```
diris %>% filter(Sepal.Length == 4.3 | Species != "setosa")
```

```
## # A tibble: 101 x 5
##    Sepal.Length Sepal.Width Petal.Length Petal.Width Species
##           <dbl>       <dbl>        <dbl>       <dbl> <chr>
## 1           4.3         3            1.1         0.1 setosa
## 2           4.9         2.5          4.5         1.7 virginica
## 3           4.9         2.4          3.3         1   versicolor
## 4           5           2.3          3.3         1   versicolor
## 5           5           2            3.5         1   versicolor
## # ... with 96 more rows
```

If we seek multiple matches in a single column, the %in% operator may be used along with a list of interest. This allows us to separate the coding that indicates our choice of interest and then filter based on interest:

```
interest <- c(4.3, 4.4)
diris %>% filter(Sepal.Length %in% interest)
```

```
## # A tibble: 4 x 5
##    Sepal.Length Sepal.Width Petal.Length Petal.Width Species
##           <dbl>       <dbl>        <dbl>       <dbl> <chr>
## 1           4.3         3            1.1         0.1 setosa
## 2           4.4         3.2          1.3         0.2 setosa
## 3           4.4         3            1.3         0.2 setosa
## 4           4.4         2.9          1.4         0.2 setosa
```

Alternately, if we wish to exclude any rows having certain characteristics, that may be done via the not in or %nin% operator. This operator can be useful to remove outliers or erroneous data points as well:

```
diris %>% filter(Sepal.Length %nin% interest)
```

```
## # A tibble: 146 x 5
##    Sepal.Length Sepal.Width Petal.Length Petal.Width Species
##           <dbl>       <dbl>        <dbl>       <dbl> <chr>
## 1           4.5         2.3          1.3         0.3 setosa
## 2           4.6         3.6          1           0.2 setosa
## 3           4.6         3.4          1.4         0.3 setosa
## 4           4.6         3.2          1.4         0.2 setosa
## 5           4.6         3.1          1.5         0.2 setosa
## # ... with 141 more rows
```

The last filter example we show uses several arguments. Using multiple arguments is not required, but can be a helpful way to break code into human-readable chunks. Multiple arguments, separated by commas, are logically equivalent to &:

```
diris %>% filter(Sepal.Length == 4.3 | Species != "setosa",
                 Petal.Width < 1.5)
```

```
## # A tibble: 37 x 5
##    Sepal.Length Sepal.Width Petal.Length Petal.Width Species
##           <dbl>       <dbl>        <dbl>       <dbl> <chr>
## 1           4.3         3            1.1         0.1 setosa
## 2           4.9         2.4          3.3         1   versicolor
## 3           5           2.3          3.3         1   versicolor
## 4           5           2            3.5         1   versicolor
## 5           5.1         2.5          3           1.1 versicolor
## # ... with 32 more rows
```

So far, we selected rows from all columns. If we wish to select only specific columns, the select() function takes a database as its first formal and column information in the rest of the formals. For named columns in your table structure, selecting one or more columns is as simple as listing the column names. It is also possible to select by position. Contrast these named vs. position column arguments and also recognize both can make the column selections in any order needed:

```
diris %>% select(Sepal.Length, Sepal.Width)
```

```
## # A tibble: 150 x 2
##    Sepal.Length Sepal.Width
##           <dbl>       <dbl>
## 1          4.3         3
## 2          4.4         3.2
## 3          4.4         3
## 4          4.4         2.9
## 5          4.5         2.3
## # ... with 145 more rows
```

```
diris %>% select(1, 5, 2)
```

```
## # A tibble: 150 x 3
##    Sepal.Length Species Sepal.Width
##           <dbl> <chr>         <dbl>
## 1          4.3 setosa          3
## 2          4.4 setosa          3.2
## 3          4.4 setosa          3
## 4          4.4 setosa          2.9
## 5          4.5 setosa          2.3
## # ... with 145 more rows
```

So far, specific columns are directly coded. What if the columns need to be variable depending on the program or the user's needs? In that case, character string references are desirable, as strings easily convert to variables. Multiple variable names as strings can be passed as separate arguments:

```
diris %>% select("Sepal.Length", "Sepal.Width")
```

```
## # A tibble: 150 x 2
##    Sepal.Length Sepal.Width
##           <dbl>       <dbl>
## 1          4.3         3
## 2          4.4         3.2
## 3          4.4         3
```

```
## 4              4.4              2.9
## 5              4.5              2.3
## # ... with 145 more rows
```

These can be placed in a variable, such as colVar1, which may also contain more than one column name:

```
colVar1 <- c("Sepal.Length", "Sepal.Width")
diris %>% select(all_of(colVar1))
```

```
## # A tibble: 150 x 2
##    Sepal.Length Sepal.Width
##           <dbl>       <dbl>
## 1           4.3           3
## 2           4.4         3.2
## 3           4.4           3
## 4           4.4         2.9
## 5           4.5         2.3
## # ... with 145 more rows
```

Additionally, select() has *helper* functions, which allow more esoteric selections to occur. An example of such a helper function is any_of(), which attempts to make as many matches as possible between the colVar2 set and the column names of the diris set. The names of these helper functions have changed a fair bit recently, so even if you are familiar with helper functions, it can help to briefly review them. Note the last entry of colVar2 is not a real entry. All the same, courtesy of the helper function used, the correct result occurs:

```
colVar2 <- c("Sepal.Length", "Sepal.Width", "FAKE!")
diris %>% select(any_of(colVar2))
```

```
## # A tibble: 150 x 2
##    Sepal.Length Sepal.Width
##           <dbl>       <dbl>
## 1           4.3           3
## 2           4.4         3.2
## 3           4.4           3
## 4           4.4         2.9
## 5           4.5         2.3
## # ... with 145 more rows
```

So far, all our work returns another `tibble` as output. Extracting the raw data as a *vector* uses several techniques. If we know the column name containing the data we want, it is easy to type the name directly. Similarly, if our column is stored in a variable, the variable may be called (be sure to only ask for one column at a time!). Regardless of the method, the console prints the same data. Notice these data are vectors, not `tibble` structures. Also notice we wrapped the functions in the head() function to save space:

```
head(diris[["Sepal.Length"]]) # easy if you have a R variable
```

```
## [1] 4.3 4.4 4.4 4.4 4.5 4.6
```

```
head(diris[[colVar1[1]]]) #easy if you know position
```

```
## [1] 4.3 4.4 4.4 4.4 4.5 4.6
```

```
head(diris$Sepal.Length) # easy to type
```

```
## [1] 4.3 4.4 4.4 4.4 4.5 4.6
```

Excluding a column or columns uses negation. Again, the column(s) to be excluded may be referenced via position or named explicitly:

```
diris %>% select(-1)
```

```
## # A tibble: 150 x 4
##    Sepal.Width Petal.Length Petal.Width Species
##          <dbl>        <dbl>       <dbl> <chr>
## 1            3          1.1         0.1 setosa
## 2          3.2          1.3         0.2 setosa
## 3            3          1.3         0.2 setosa
## 4          2.9          1.4         0.2 setosa
## 5          2.3          1.3         0.3 setosa
## # ... with 145 more rows
```

```
diris %>% select(-Sepal.Length, -Petal.Width)
```

```
## # A tibble: 150 x 3
##    Sepal.Width Petal.Length Species
##          <dbl>        <dbl> <chr>
## 1            3          1.1 setosa
## 2          3.2          1.3 setosa
```

```
## 3          3              1.3 setosa
## 4          2.9            1.4 setosa
## 5          2.3            1.3 setosa
## # ... with 145 more rows
```

When variable exclusion is needed, we set a variable to hold the vector, and we may simply negate that variable. Alternately, we may use negation with the helper functions:

```
diris %>% select(-colVar1)
```

```
## Note: Using an external vector in selections is ambiguous.
## i Use 'all_of(colVar1)' instead of 'colVar1' to silence this message.
## i See <https://tidyselect.r-lib.org/reference/faq-external-vector.html>.
## This message is displayed once per session.
```

```
## # A tibble: 150 x 3
##    Petal.Length Petal.Width Species
##           <dbl>       <dbl> <chr>
## 1          1.1         0.1 setosa
## 2          1.3         0.2 setosa
## 3          1.3         0.2 setosa
## 4          1.4         0.2 setosa
## 5          1.3         0.3 setosa
## # ... with 145 more rows
```

```
diris %>% select(-any_of(colVar2))
```

```
## # A tibble: 150 x 3
##    Petal.Length Petal.Width Species
##           <dbl>       <dbl> <chr>
## 1          1.1         0.1 setosa
## 2          1.3         0.2 setosa
## 3          1.3         0.2 setosa
## 4          1.4         0.2 setosa
## 5          1.3         0.3 setosa
## # ... with 145 more rows
```

Having now used the any_of() helper function a couple of times, we look briefly at three more helper functions. These three are starts_with(), ends_with(), and contains() and allow us to choose variables without typing out an entire column name. They default to ignoring letter cases, although that is an option that can be set to FALSE. One benefit of the tidyverse approach of making focused functions named after verbs that accomplish one specific task is it is easy to tell what a function does from its name:

```
diris %>% select(starts_with("sepal"))
```

```
## # A tibble: 150 x 2
##    Sepal.Length Sepal.Width
##           <dbl>       <dbl>
## 1          4.3           3
## 2          4.4         3.2
## 3          4.4           3
## 4          4.4         2.9
## 5          4.5         2.3
## # ... with 145 more rows
```

```
diris %>% select(starts_with("sepal", ignore.case = FALSE))
```

```
## # A tibble: 150 x 0
```

The last helper function we mention is matches(). This function uses regular expressions, such as the one shown, to match specific patterns of columns. A full treatment of regular expressions is not in order for this text. However, in the following expression, the pattern match is one that starts with at least one alphabetic character, has a single punctuation mark, and then has at least one alphabetic character after the punctuation mark:

```
diris %>% select(matches("^[[:alpha:]]+[[:punct:]]{1}[[:alpha:]]+"))
```

```
## # A tibble: 150 x 4
##    Sepal.Length Sepal.Width Petal.Length Petal.Width
##           <dbl>       <dbl>        <dbl>       <dbl>
## 1          4.3           3          1.1         0.1
## 2          4.4         3.2          1.3         0.2
## 3          4.4           3          1.3         0.2
```

```
## 4            4.4         2.9         1.4         0.2
## 5            4.5         2.3         1.3         0.3
## # ... with 145 more rows
```

With this last bit of code, our journey through tidyverse subsetting ends. We turn our attention to variable renaming and ordering.

9.3 Variable Renaming and Ordering

It is a regrettably routine part of data hygiene that column names require renaming or reordering. What made sense (or is "legal") in one database may not work in another data bank. Many proprietary data management systems have fixed names that are unwieldy. Viewing the names associated with a tidyverse object requires the usual names() or colnames() function called on our data set diris. Either way, we get the same results with the current column order:

```
diris %>% names()

## [1] "Sepal.Length" "Sepal.Width" "Petal.Length" "Petal.Width"
## [5] "Species"

diris %>% colnames()

## [1] "Sepal.Length" "Sepal.Width" "Petal.Length" "Petal.Width"
## [5] "Species"
```

To change a name, call the rename() function on the data object and set the new name in the second formal by using the following format: new name = old name. Multiple columns can be renamed at once by passing them as additional arguments (e.g., in the third or fourth formal). Confirm the change took place by comparing the results of name() with the old results. In the following code, we remove the full stop between Sepal and Length in our data:

```
diris <- diris %>% rename(SepalLength = Sepal.Length)
diris %>% names()

## [1] "SepalLength" "Sepal.Width" "Petal.Length" "Petal.Width"
## [5] "Species"
```

To reorder the columns to pair the lengths and the widths together, simply use the select() function and reassign a name to your data object. Be sure to remember we renamed SepalLength to no longer have a full stop in the middle!

```
diris <- diris %>% select(
                SepalLength,
                Petal.Length,
                Sepal.Width,
                Petal.Width,
                Species)

diris

## # A tibble: 150 x 5
##    SepalLength Petal.Length Sepal.Width Petal.Width Species
##          <dbl>        <dbl>       <dbl>       <dbl> <chr>
## 1          4.3          1.1           3         0.1 setosa
## 2          4.4          1.3         3.2         0.2 setosa
## 3          4.4          1.3           3         0.2 setosa
## 4          4.4          1.4         2.9         0.2 setosa
## 5          4.5          1.3         2.3         0.3 setosa
## # ... with 145 more rows
```

Earlier, you saw numeric select calls could reorder columns, so we close out this section with one last reordering example. Here, we use a variable to hold numeric positions, although we could just as well use character names in place of numbers. Then we use the select() function to reorganize our columns back to their original order:

```
colVar3 <- c(1, 3, 2, 4, 5)
diris %>% select(all_of(colVar3))

## # A tibble: 150 x 5
##    SepalLength Sepal.Width Petal.Length Petal.Width Species
##          <dbl>       <dbl>        <dbl>       <dbl> <chr>
## 1          4.3           3          1.1         0.1 setosa
## 2          4.4         3.2          1.3         0.2 setosa
## 3          4.4           3          1.3         0.2 setosa
## 4          4.4         2.9          1.4         0.2 setosa
## 5          4.5         2.3          1.3         0.3 setosa
## # ... with 145 more rows
```

9.4 Computing on Data and Creating Variables

Now that you have a good sense of how to manage original data in `tidyverse`, we focus on how to create new data from old data. For simplicity's sake, we re-create our `diris` data set from the original `iris` data set. Then we use the `mutate()` function to create or replace columns. This function takes the usual first argument, and the latter formals express what is done to create new columns. Whether it is one or many, simply create expressions that let tidyverse know what to append to the `tibble`. Notice here we do not reassign this mutation back to `diris`. Thus, this mutation is not saved. In exploratory analysis, it can be convenient to run such mutations from the console to understand features of data:

```
diris <- as_tibble(iris)
diris %>% mutate(V0 = 0, X1 = 1L, X2 = 2L)
```

```
## # A tibble: 150 x 8
##    Sepal.Length Sepal.Width Petal.Length Petal.Width Species    V0
##           <dbl>       <dbl>        <dbl>       <dbl> <chr>   <dbl>
## 1           5.1         3.5          1.4         0.2 setosa      0
## 2           4.9         3            1.4         0.2 setosa      0
## 3           4.7         3.2          1.3         0.2 setosa      0
## 4           4.6         3.1          1.5         0.2 setosa      0
## 5           5           3.6          1.4         0.2 setosa      0
## # ... with 145 more rows, and 2 more variables: X1 <int>, X2 <int>
```

Of course, new data might not be about creating wholly new columns. Instead, we may create new calculated columns based on existing information. Perhaps we want a new column named V that is the multiplication of `Petal.Length` and `Petal.Width`:

```
diris <- mutate(diris, V = Petal.Length * Petal.Width)
diris
```

```
## # A tibble: 150 x 6
##    Sepal.Length Sepal.Width Petal.Length Petal.Width Species      V
##           <dbl>       <dbl>        <dbl>       <dbl> <chr>    <dbl>
## 1           5.1         3.5          1.4         0.2 setosa   0.280
## 2           4.9         3            1.4         0.2 setosa   0.280
## 3           4.7         3.2          1.3         0.2 setosa   0.26
```

```
## 4            4.6            3.1            1.5            0.2 setosa  0.3
## 5              5            3.6            1.4            0.2 setosa  0.280
## # ... with 145 more rows
```

More exotic calculations are performed only when certain conditions occur. In this case, if_else() takes three arguments. The first is the logical condition on which to test the if/else statement. The second and third formals indicate what to do on the if or the else. Additionally, mutate() allows us not only to create a column with data but also to change existing column data. Thus, if it was not enough in our first pass to create a column V2 that took on a multiplicative relationship between petal widths and lengths for setosa species, we could extend that to a square root operation for virginica. To make sure we see a sample of each species, we use the slice() function to look at just three specific rows:

```
diris <- diris %>% mutate(V2 = if_else(Species == "setosa",
                                    Petal.Length * Petal.Width, NA_real_))
diris <- diris %>% mutate(V2 = if_else(Species == "virginica",
                                    sqrt(Petal.Length * Petal.Width), V2))
diris %>% slice(c(1, 51, 101))
```

```
## # A tibble: 3 x 7
##   Sepal.Length Sepal.Width Petal.Length Petal.Width Species       V
##          <dbl>       <dbl>        <dbl>       <dbl> <chr>     <dbl>
## 1          5.1         3.5          1.4         0.2 setosa    0.280
## 2            7         3.2          4.7         1.4 versic~   6.58
## 3          6.3         3.3            6         2.5 virgin~   15
## # ... with 1 more variable: V2 <dbl>
```

It is not necessary to create a new column; it is possible to simply calculate information on a table by using the summarise() function. The first formal, as always, tells the function which data to use (although that is hidden by virtue of the pipe), while the rest are used to hold individual calculations. Note that one or many calculations might be coded. Here, we demonstrate two such calculations for arithmetic mean() and standard deviation sd():

```
## computing in summarise allows multiple operations
diris %>% summarise(M = mean(Sepal.Width),
                    SD = sd(Sepal.Width))
```

```
## # A tibble: 1 x 2
##       M     SD
##   <dbl> <dbl>
## 1  3.06 0.436
```

Suppose we want to take our data, use only the `virginica` species, and calculate the mean of the sepal widths. The result should be a `tibble` of just one variable, the mean we seek. Here is where the pipe operator starts to show its strength, because this code is very readable. Indeed, the code may well make sense even to non-programming types:

```
diris %>%
  filter(Species == "virginica") %>%
  summarise(M = mean(Sepal.Width))
```

```
## # A tibble: 1 x 1
##       M
##   <dbl>
## 1  2.97
```

This idea of taking it one step at a time can be modified to manage all sorts of data manipulations. Thinking back to our study of `data.table`, the third formal of that structure, by `=`, allowed us to organize by elements of a column. For tidyverse, we use `group_by()` on a column name to accomplish the same task. Expanding our mean example to all three species, we now have a two-column tibble with one row for each species. We could also get means for more than one of our original columns:

```
diris %>%
  group_by(Species) %>%
  summarise(
    mSW = mean(Sepal.Width),
    mPW = mean(Petal.Width))
```

```
## # A tibble: 3 x 3
##   Species       mSW   mPW
##   <chr>       <dbl> <dbl>
## 1 setosa       3.43 0.246
## 2 versicolor   2.77  1.33
## 3 virginica    2.97  2.03
```

This chaining structure, just like the data table chaining structure, can be done indefinitely (or close to it). To find a correlation in the preceding data, we simply add one more step to the end of our chain with the cor() function:

```
diris %>%
  group_by(Species) %>%
  summarise(
    mSW = mean(Sepal.Width),
    mPW = mean(Petal.Width)) %>%
  summarise(r = cor(mSW, mPW))

## # A tibble: 1 x 1
##         r
##     <dbl>
## 1 -0.759
```

Variables can create other variables as well. We create a new boolean variable based on whether a petal width is greater or smaller than the median petal width for that species:

```
diris <- diris %>%
  group_by(Species) %>%
  mutate(MedPW = Petal.Width > median(Petal.Width))
diris

## # A tibble: 150 x 8
## # Groups: Species [3]
##    Sepal.Length Sepal.Width Petal.Length Petal.Width Species      V
##           <dbl>       <dbl>        <dbl>       <dbl> <chr>    <dbl>
## 1           5.1         3.5          1.4         0.2 setosa   0.280
## 2           4.9         3            1.4         0.2 setosa   0.280
## 3           4.7         3.2          1.3         0.2 setosa   0.26
## 4           4.6         3.1          1.5         0.2 setosa   0.3
## 5           5           3.6          1.4         0.2 setosa   0.280
## # ... with 145 more rows, and 2 more variables: V2 <dbl>, MedPW <lgl>
```

It is also easy to group by multiple variables. Grouping by both species and median petal width booleans, we recalculate means for sepal and petal widths:

```
diris %>%
  group_by(Species, MedPW) %>%
  summarise(
    mSW = mean(Sepal.Width),
    mPW = mean(Petal.Width))
```

```
## # A tibble: 6 x 4
## # Groups:   Species [3]
##   Species    MedPW   mSW   mPW
##   <chr>      <lgl> <dbl> <dbl>
## 1 setosa     FALSE  3.38 0.185
## 2 setosa     TRUE   3.54 0.375
## 3 versicolor FALSE  2.62 1.19
## 4 versicolor TRUE   2.96 1.50
## 5 virginica  FALSE  2.84 1.81
## 6 virginica  TRUE   3.13 2.27
```

Again, it is always possible to chain on more operations:

```
diris %>%
  group_by(Species, MedPW) %>%
  summarise(
    mSW = mean(Sepal.Width),
    mPW = mean(Petal.Width)) %>%
  group_by(MedPW) %>%
  summarise(r = cor(mSW, mPW))
## # A tibble: 2 x 2
##   MedPW     r
##   <lgl> <dbl>
## 1 FALSE -0.778
## 2 TRUE  -0.760
```

9.5 Merging and Reshaping Data

Without repeating what we said about merging and reshaping data in Chapter 7, we note the function merge() still works, after a fashion. Since tibbles are built on data frames, it works. However, when this is done, they lose their tibble status. Thus, we also introduce the *join() functions, which are much closer to some of the more traditional database languages.

We first run a bit of code to make four separate data objects. Please note this is not likely to be needed in real life, as our data usually comes from different databases:

```
#Data Set One
diris <- diris %>%
  group_by(Species) %>%
  select(Species, Sepal.Width) %>%
  slice(1:3)
diris

## # A tibble: 9 x 2
## # Groups:   Species [3]
##    Species     Sepal.Width
##    <chr>             <dbl>
## 1 setosa              3.5
## 2 setosa              3
## 3 setosa              3.2
## 4 versicolor          3.2
## 5 versicolor          3.2
## 6 versicolor          3.1
## 7 virginica           3.3
## 8 virginica           2.7
## 9 virginica           3

#Data Set Two
diris2 <- diris %>% slice(c(1, 4, 7))
diris2
## # A tibble: 3 x 2
## # Groups:   Species [3]
##    Species     Sepal.Width
```

```
##   <chr>              <dbl>
## 1 setosa               3.5
## 2 versicolor           3.2
## 3 virginica            3.3

#Data Set Three
dalt1 <- tibble(
  Species = c("setosa", "setosa", "versicolor", "versicolor",
              "virginica", "virginica", "other", "other"),
  Type = c("wide", "wide", "wide", "wide",
           "narrow", "narrow", "moderate", "moderate"),
  MedPW = c(TRUE, FALSE, TRUE, FALSE, TRUE, FALSE, TRUE, FALSE))
dalt1
## # A tibble: 8 x 3
##   Species    Type     MedPW
##   <chr>      <chr>    <lgl>
## 1 setosa     wide     TRUE
## 2 setosa     wide     FALSE
## 3 versicolor wide     TRUE
## 4 versicolor wide     FALSE
## 5 virginica  narrow   TRUE
## 6 virginica  narrow   FALSE
## 7 other      moderate TRUE
## 8 other      moderate FALSE

#Data Set Four
dalt2 <- dalt1 %>% slice(c(1, 3))
dalt2

## # A tibble: 2 x 3
##   Species    Type  MedPW
##   <chr>      <chr> <lgl>
## 1 setosa     wide  TRUE
## 2 versicolor wide  TRUE
```

The merge() function operates on two data sets. For one-to-one data such as diris2 and dalt2, the merge by itself keeps only the rows that match both data sets. We can see that for these two objects, virginica row in diris2 is dropped:

```
diris2 %>% merge(dalt2)
```

```
##       Species Sepal.Width Type MedPW
## 1      setosa         3.5 wide  TRUE
## 2 versicolor         3.2 wide  TRUE
```

If we wish to keep all three rows of diris2, then there is some missing data after the merge. Now, in this case, since we arbitrarily choose diris2 to be our first formal, all.x = TRUE keeps all of the diris2 rows:

```
diris2 %>% merge(dalt2, all.x = TRUE)
```

```
##       Species Sepal.Width Type MedPW
## 1      setosa         3.5 wide  TRUE
## 2 versicolor         3.2 wide  TRUE
## 3  virginica         3.3 <NA>    NA
```

It is also possible to set the second value, or all.y = TRUE, although in this case that has no noticeable difference. Depending on how good a match two particular tables are, there might be a different pattern of missing variables. Similarly, it is also possible to say that we want all rows from all tables. This has the potential to give the most missing data locations; in our example, such a merge would be identical to our three rows shown in the preceding example:

```
diris2 %>% merge(dalt2, all.y = TRUE)
```

```
##       Species Sepal.Width Type MedPW
## 1      setosa         3.5 wide  TRUE
## 2 versicolor         3.2 wide  TRUE
```

```
diris2 %>% merge(dalt2, all = TRUE)
```

```
##       Species Sepal.Width Type MedPW
## 1      setosa         3.5 wide  TRUE
## 2 versicolor         3.2 wide  TRUE
## 3  virginica         3.3 <NA>    NA
```

As mentioned, these are not tibbles. Now, we may wrap them in as_tibble() if we wish to merge and coerce back to a tibble. However, tidyverse also operates on database-style objects, as you will see in Chapter 10. In particular, *Structured Query Language* (SQL) can be accessed through dplyr. SQL has joins, and we consider four types of joins now.

All rows of the first object are returned in a left_join() as well as all columns of both the first and second objects. Because this is a dplyr function call, we get a tibble class returned output:

```
diris2 %>% left_join(dalt2)

## Joining, by = "Species"

## # A tibble: 3 x 4
## # Groups: Species [3]
##    Species   Sepal.Width Type  MedPW
##    <chr>           <dbl> <chr> <lgl>
## 1 setosa            3.5  wide  TRUE
## 2 versicolor        3.2  wide  TRUE
## 3 virginica         3.3  <NA>  NA
```

A right_join() returns a tibble with the columns of both objects and all rows from the *second* object. Many charts on the Internet claim to speak the truth about how joins work. In our experience, the accuracy of those explanations can be hit or miss. At the very least, most people find many such graphs confusing. Our advice is to grab several small data tables or cook up your own as we did here. Then start joining things. This is the learning to ride a bicycle approach. Rather than discuss the physics of bi-wheel movement, simply try often, and do not fear some scrapped knees along the way:

```
diris2 %>% right_join(dalt2)

## Joining, by = "Species"

## # A tibble: 2 x 4
## # Groups:   Species [2]
##    Species    Sepal.Width Type  MedPW
##    <chr>            <dbl> <chr> <lgl>
## 1 setosa             3.5  wide  TRUE
## 2 versicolor         3.2  wide  TRUE
```

The inner_join() function in this case yields the same results as the previous right join. However, only rows matching up in both objects are returned. Perhaps the weirdest join is the anti_join() which returns all rows in the first database not matching values in the second data object. As for columns, it returns only the first object's columns:

```
diris2 %>% inner_join(dalt2)

## Joining, by = "Species"

## # A tibble: 2 x 4
## # Groups: Species [2]
##   Species   Sepal.Width Type  MedPW
##   <chr>           <dbl> <chr> <lgl>
## 1 setosa            3.5 wide  TRUE
## 2 versicolor        3.2 wide  TRUE

diris2 %>% anti_join(dalt2)

## Joining, by = "Species"

## # A tibble: 1 x 2
## # Groups:   Species [1]
##   Species   Sepal.Width
##   <chr>           <dbl>
## 1 virginica         3.3
```

Between joins and merges, there are plenty of options for combining data into a single object. Remember, your goal is to eventually get just the data needed for analysis. Sometimes it makes sense to first combine one or more collections of data into a single table and from there use the techniques from earlier in this chapter to reduce that to your desired data set. Of course, if merges are more familiar, they often work just fine. If you truly require tidyverse, as_tibble() always works as a wrapper function to coerce your object back to what it should be. After all, data.table and tidyverse's tibbles are both extensions of data frames.

Our last goal is reshaping, which works better if there is a unique ID per row. We start by refreshing our data set one more time. From there, we introduce n(), which is a special way to refer to the number of rows, much like .N is in data tables:

```
diris <- as_tibble(iris)
diris <- diris %>% mutate(ID = 1:n())
diris
```

```
## # A tibble: 150 x 6
##     Sepal.Length Sepal.Width Petal.Length Petal.Width Species     ID
##            <dbl>       <dbl>        <dbl>       <dbl> <chr>    <int>
## 1           5.1         3.5          1.4         0.2 setosa       1
## 2           4.9         3            1.4         0.2 setosa       2
## 3           4.7         3.2          1.3         0.2 setosa       3
## 4           4.6         3.1          1.5         0.2 setosa       4
## 5           5           3.6          1.4         0.2 setosa       5
## # ... with 145 more rows
```

To take our data long, we use the `tidyr` function `pivot_longer()`. This function takes a list of columns from our data which are to be modified into the new format. In this case, it is our Sepal and Petal variables. The `names_to` argument uses a special command, `.value`, to signal we want the new column(s) to contain the values from the old columns. Contrastingly, the next element Type will contain the information that signals which column the data came from. Because we have *two* entries in `names_to`, our `names_pattern` expects two parts to it. Using regular expressions, we signal the variable pieces with | for logical *or*:

```
diris.long <-
  diris %>% pivot_longer(
    cols = c(Sepal.Length, Sepal.Width, Petal.Length, Petal.Width),
    names_to = c(".value", "Type"),
    names_pattern = "(Sepal|Petal).(Width|Length)"
  )
diris.long
```

```
## # A tibble: 300 x 5
##     Species      ID Type    Sepal Petal
##     <chr>    <int> <chr>   <dbl> <dbl>
## 1 setosa       1 Length    5.1   1.4
## 2 setosa       1 Width     3.5   0.2
## 3 setosa       2 Length    4.9   1.4
```

```
## 4 setosa      2 Width    3     0.2
## 5 setosa      3 Length   4.7   1.3
## # ... with 295 more rows
```

We took wider data and made it long. We now reverse the process by using `pivot_wider()`. The `names_from` signals where the wider column names will be from. The separator defaults to `_`, and thus we change it to a full stop. Lastly, `values_from` contains the columns with our data values:

```
diris.wide <- diris.long %>% pivot_wider(
  names_from = Type,
  names_sep = ".",
  values_from = c(Sepal, Petal)
  )

diris.wide
```

```
## # A tibble: 150 x 6
##    Species    ID Sepal.Length Sepal.Width Petal.Length Petal.Width
##    <chr>   <int>        <dbl>       <dbl>        <dbl>       <dbl>
## 1 setosa      1          5.1         3.5          1.4         0.2
## 2 setosa      2          4.9         3            1.4         0.2
## 3 setosa      3          4.7         3.2          1.3         0.2
## 4 setosa      4          4.6         3.1          1.5         0.2
## 5 setosa      5          5           3.6          1.4         0.2
## # ... with 145 more rows
```

9.6 Summary

The `dplyr`, `tibble`, and `tidyr` packages can make working with data more human readable. By separating what we do to data from the data itself, we can learn a comparatively limited number of manipulations applicable to any data. Behind the scenes, `tidyverse` translates those function calls to modify the data. In the next chapter, you will see how this allows us to quickly access big data. Table 9-1 describes the key functions used in this chapter.

Table 9-1. *Listing of key functions described in this chapter and summary of what they do*

Function	What It Does
`as_tibble()`	Converts a data object (e.g., a data frame) to a tibble.
`arrange()`	Arranges a named column in increasing order, or descending order if `desc()` is used on the column name.
`distinct()`	This is `dplyr`'s answer for `unique()`. It gives just those rows that are not the same.
`slice()`	Slices out the rows asked for from a tibble.
`filter()`	Filters out rows based on matched characteristics.
`select()`	Selects specific columns based on name. Functionality can be expanded by using the helper functions: `one_of()`, `starts_with()`, `ends_with()`, `contains()`, `matches()`, `num_range()`, and `everything()`.
`rename()`	Renames columns.
`mutate()`	Allows for changes or mutations to existing tibbles. Adding/deleting/modifying column data is a popular use.
`summarise()`	Creates a new tibble based on calculations from an original tibble.
`%>%`	Piping avoids function composition and allows for cleaner code to chain up multiple functions.
`group_by()`	Allows grouping by columns.
`left_join()`	Joins are an SQL way of dealing with data. A left join keeps the first object's rows.
`right_join()`	Keeps the second object's rows.
`inner_join()`	Returns all columns of both variables and all rows from the first that have matches in the second data object.
`anti_join()`	Returns just the unique, unlinkable part of the first data object. This is quite helpful if the second object is meant to extend information on the first, as it gives a fast way to see what is left.
`pivot_longer()`	Takes wider data and melts it longer.
`pivot_wider()`	Takes longer data and casts it wider.

CHAPTER 10

Reading Big Data

With your ability to understand *manipulating* data, it is now worth taking a moment to consider where data live. R works in memory, random access memory (RAM), not hard drives. A quick check of your system settings should reveal the amount of memory you have. We, the authors, use between 16 and 128 GB in our real-world systems, with the larger number being an expensive yet convenient habit. R cannot analyze data larger than available RAM. This becomes important when considering the ever larger file sizes one encounters in modern research. Too much data in RAM would stop R and your computer system from working. Thus one must take care to ensure just the right data arrive in RAM.

How do data arrive in RAM? Smaller data are often found in comma-separated values (CSV) files or files easily converted to such. However, larger data tend to live in other places called databases. While most databases have the capability to export files to CSV, this is not always convenient. Being able to directly access data from a database and (potentially) being able to write data out of R into a database are key skills to successfully analyzing bigger data.

To reduce complexity, in this chapter, we install and host a database server on a local machine. That said, most cloud compute solutions have options to host databases on the Web. Other than the installation/setup process, the mechanics of connecting and interfacing R and a database are the same regardless of the specific location of your database.

The database we will use is PostgreSQL [?] which is both open source and popular. Highly advanced, this database has over 30 types of data it understands internally. We recommend *Beginning Databases with PostgreSQL* by Neil Matthew and Richard Stones (Apress, 2005). We will walk you through installing a practice version on your local machine. Our goal is to get you using R on data inside this database as soon as possible. More advanced database operations most likely ought not to be done in R.

© Matt Wiley and Joshua F. Wiley 2020
M. Wiley and J. F. Wiley, *Advanced R 4 Data Programming and the Cloud*,
https://doi.org/10.1007/978-1-4842-5973-3_10

PostgreSQL incorporates many highly useful database features including data reliability measures and various levels of user access control. While the predominant focus of this text is R, we will *briefly* delve into just enough database management information (safely contained in a stand-alone section) for readers who are wishing to get started on their own database. Nevertheless, our focus is on practicing enough key skills that you are able to access and analyze database data.

This chapter requires the following packages: DBI [17], RPostgreSQL [7], data.table [9], keyring [8], and JWileymisc [36].

```
library(RPostgreSQL)
library(DBI)
library(data.table)
library(JWileymisc)
library(keyring)

options(width = 70,
        stringsAsFactors = FALSE,
        datatable.print.nrows = 20, ## if over 20 rows
        datatable.print.topn = 3, ## print first and last 3
        digits = 2) ## reduce digits printed by R
```

10.1 Installing PostgreSQL on Windows

Visit the PostgreSQL website www.postgresql.org/download/windows/ and choose one of the installer options. We chose the installer for 64-bit architecture and version 12.2 (which was the latest release as of this writing). After downloading, you should have a file approximately named postgresql-12.2-2-windows-x64.exe, which is your installation program. Of course, your version number likely is higher than 12.2. Most likely a more recent version will not make a major change in your ability to follow along with this chapter.

Go ahead and install the program, clicking the *Next* button as needed and accepting the default settings for the installation directory. The installer asks for a password for the database super user named postgres, which we set to advancedr. If you select a different password, please be sure to record it securely. Incidentally, this also creates a database with the same name as the super user. PostgreSQL defaults to port 5432, and if that is not showing as the default option, you likely have another version of PostgreSQL living on your computer. Clicking *Next* a few more times, accepting the defaults, should lead you to the installation event. You may be asked on the final screen about *Stack Builder*; it is **not** required, and you may *uncheck* that option before clicking *Finish*.

At this point, PostgreSQL's installation is over, and it is running on your computer. If you are familiar with this database, you could add data to it directly (and likely would not have needed this primer). We add data via R and quickly get to accessing subsets of that data.

10.2 Interfacing R and PostgreSQL

Our first task is connecting to our database. This uses the `DBI` package to build a link to the database we just installed. The `dbDriver()` function is a `DBI` function that sets the default interface instructions. In this case, we need a PostgreSQL driver, which is provided via `RPostgreSQL`. This is worth noting, because `DBI` is a wrapper that works for several database types. There are other options besides PostgreSQL, and many of these can be accessed via `DBI` functions. Thus, the majority of this chapter remains true regardless of database. For readers who must interface with several different databases, this can be quite convenient to reduce mental cognitive load by staying in the RStudio environment and predominantly using R functions. In fact, it can be quite seamless to transition between different databases and not really notice. The result of `dbDriver()` is assigned to a variable, which we call `drv`.

With the correct mechanism or driver for interfacing, it is time to connect specifically to our database. When we installed PostgreSQL, we created a super user named `postgres` and a password of `advancedr`. Additionally, the install process created a first database with the same name as our super user, namely, `postgres`. This super user account has full privileges to all tables in the database. We will discuss user privileges more in depth later.

To connect to this database named `postgres`, we will use our driver and must specify the host location. Because we installed the server on our local machine, we will use `host = "localhost"`. However, if you had chosen instead to use a cloud service (which we cover more in depth in later chapters), you would simply put the web address of your server inside the quotes. It is also important to set the port, which you will recall was part of the install process. Those need to match, and on the install, we stayed with the default port of `5432`. If you were using a cloud service, your port might also be different. Lastly, we use the username and password we set in the install process.

All of this is used in the dbConnect() function to build a connection to our database server. In this case, we name that conSuper for "connection for super user." This will be the connection created with our super user's username and password.

The last step of our *first* connection process is to use the dbGetInfo() to confirm we have successfully connected to our database. It is not required to use this function every time you connect. Rather, we use it here as verification that all systems are go. Provided your output mostly matches our output, you know your install process worked:

```
drv <- dbDriver("PostgreSQL")

conSuper <- dbConnect( drv,
                 dbname = "postgres",
                 host = "localhost",
                 port = 5432,
                 user = "postgres",
                 password = "advancedr"
                 )

dbGetInfo(conSuper)
## $host
## [1] "localhost"
##
## $port
## [1] "5432"
##
## $user
## [1] "postgres"
##
## $dbname
## [1] "postgres"
##
## $serverVersion
## [1] "12.0.2"
##
## $protocolVersion
## [1] 3
```

```
##
## $backendPId
## [1] 23164
##
## $rsId
## list()
```

The connection we just established links our R instance to our database. Right now, this database happens to be empty. If your role is that of an analyst, you may have a real-life database already full of data. On the other hand, you may have data of your own you are seeking to move into formal database systems. In either case, we explore both pushing data into our server and pulling data from our server.

To perform either a pull or a push, we will need to have a live connection open. However, before we are ready to explore that process, we need to briefly consider some database philosophies. Thus, for now, we close this connection. This requires two steps, because we must both disconnect from PostgreSQL and unload our driver that allows DBI to make a connection to our particular flavor of database. If successful, both dbDisconnect() and dbUnloadDriver() will return TRUE. Even if they return true, the information will persist in the *global environment*. Thus, we also use rm() to clean up our environment:

```
dbDisconnect(conSuper)
```

```
## [1] TRUE
```

```
dbUnloadDriver(drv)
```

```
## [1] TRUE
```

```
rm(conSuper)
rm(drv)
```

Provided all these steps worked, you now have a fully functional PostgreSQL database server running on your local machine. You also have the ability to connect and disconnect R to that server. Before we discuss specific techniques for using SQL and R, we briefly discuss database philosophy.

10.3 Database Philosophy

While this text is not intended to teach either databases in general nor SQL databases in particular, it is worth having a high-level concept of how such data stores work. The data base we use, PostgreSQL, is an SQL database. It belongs to the class of *relational* databases. The rough idea is every *related* set of ideas should live on a *table*. Let us consider how these tables relate for a moment.

Consider a research project of the type you might have participated in during your undergraduate years (or be designing now). Participants' names and contact information may be required. Thus, those would live on an ADDRESS table. Similarly, there may be some intake data collected about a participant (e.g., biological sex, birthdate, socioeconomic status, education level, etc.). These might live on a DEMOGRAPHICS table. Lastly, there could be regular time-series measures (e.g., stress at various times of day or hours of sleep per night) which could live in a TIMEMETRICS table.

Each of these tables is somewhat independent. The ADDRESS table requires numeric and text data; it may also not be regularly updated. Similarly, while somewhat different, the initial intake information in DEMOGRAPHICS establishes a baseline. Contrastingly, the daily or nightly time-series measures in TIMEMETRICS might be expected to change often.

What is gained by splitting these related yet distinct data? Why not simply store in a CSV file? Three main reasons are size of data, confidentiality, and reliability.

Overall, most operating systems have a limit on the size of any single file. SQL databases in general, and PostgreSQL specifically, have both data store files *and* data management processes. As such, the size on hard disk for PostgreSQL can be significantly larger (e.g., perhaps the difference between 1 GB and 64 TB). While such sizes could be accommodated via multiple CSV files, the management of those files would need to be achieved via personal effort rather than more automated database methods and standards.

Most shared folders have a limited number of permissions that can be modified. Thus, a folder of CSV files is often equally shared between all users. Contrastingly, PostgreSQL can have multiple roles and users created in a custom fashion. Thus, some users may have only read access to perhaps only some tables of data. In our example of a human subjects research study, this would allow only the principal investigator to have access to ADDRESS, while multiple researchers may need access to TIMEMETRICS. Again, much like file size, it is not that some alternative system might exist via file shares. Rather, it is more a case of "Would it be easier if this was all managed under a standard protocol?"

Furthermore, for some use cases involving frequent data writing or manipulation, not all files are able to cope with failures of data movement. The database standards that PostgreSQL implements include provisions for eventualities such as power failure or writing errors. Data stored to these higher standards are simply more likely to survive a less-than-ideal event.

What is the cost of gaining these three features? In short, every unique table must not only include the specific raw data it stores, it must also store each row of data with some sort of connecting *key*. This key allows data from table A to be linked to data in table B. In the case of our example, what would connect each table would be a `UserID`. Every table requires at least one key, and in some cases, more than one key column might be required. As an example of a table that has more than one column contributing to the key (which serves as a unique identifier for every row), consider the `TIMEMETRICS` column in our example. With multiple daily metrics per user, it would also require information about `SurveyDay` and perhaps `SurveyStartTime` to uniquely mark each row.

Because of this need for a unique key for each table, the cost of using a database is that tables may need some level of advance planning to ensure the entire database works in the correct fashion. Additionally, accessing data from multiple tables requires *joining* data from needed tables together via the keys. This does add a layer of complexity to data sourcing. That said, most large data systems are stored in such a method.

In the next section, we take simulated data from one of your authors' research projects and provision our `postgres` database with that information. This provisioning is done using the preceding philosophy and is meant to be an example or sampler rather than a full-fledged solution.

10.4 Adding Data to PostgreSQL

Using the data `aces_daily` from the `JWileymisc` package, we split the flat data into `ADDRESS`, `DEMOGRAPHICS`, `THRICEDAILYMETRICS`, and `DAILYMETRICS` tables. Because these are simulated data based on an actual study, the address table is further simulated with only one column, a faux postal code. To ensure your simulation matches ours, we use `set.seed(42020)` to ensure the data are the same. For this example, we will not actually use the pseudo-location data for any direct analysis. We also convert all table column names to lowercase. PostgreSQL can be a trifle finicky with capital letters for variables:

```
aces_daily <- as.data.table(aces_daily)
ADDRESS <- aces_daily[, .(UserID)]
ADDRESS <- unique(ADDRESS)

set.seed(42020)
ADDRESS$PostalCode <- round(rnorm(nrow(ADDRESS), mean = 4000, sd = 300), 0)
colnames(ADDRESS) <- tolower(colnames(ADDRESS))

DEMOGRAPHICS <- aces_daily[, .(UserID, Age, BornAUS, SES_1, EDU)]
DEMOGRAPHICS <- unique(DEMOGRAPHICS)
DEMOGRAPHICS[, SurrogateID := seq_len(.N)]
DEMOGRAPHICS <- DEMOGRAPHICS[, .(UserID, SurrogateID, Age, BornAUS, SES_1, EDU)]
colnames(DEMOGRAPHICS) <- tolower(colnames(DEMOGRAPHICS))

THRICEDAILYMETRICS <- aces_daily[, .(UserID, SurveyInteger,
SurveyStartTimec11, STRESS, PosAff, NegAff)]
colnames(THRICEDAILYMETRICS) <- tolower(colnames(THRICEDAILYMETRICS))

DAILYMETRICS <- aces_daily[, .(UserID, SurveyDay,
                               SOLs, WASONs, SUPPORT,
                               COPEPrb, COPEPrc, COPEExp, COPEDis)]
colnames(DAILYMETRICS) <- tolower(colnames(DAILYMETRICS))
rm(aces_daily)
```

Now that we have our sample study data, it is time to save our data in our database.

To set up a database, we must save our tables in it. Recalling our database's name (dbname = "postgres"), we use our super user credentials to connect to the database:

```
drv <- dbDriver("PostgreSQL")
conSuper <- dbConnect(drv, dbname = "postgres", host = "localhost",port = 5432,
                  user = "postgres", password = "advancedr" )
```

With the connection active, we use a new function, dbWriteTable(), to push our four tables to the SQL server. In our case, for simplicity, we make the SQL table's name match the values being written. As each table is written, a boolean is returned specifying if the write was successful. After writing our four data tables to our SQL server, we remove them from our local R environment.

One item of note: PostgreSQL table names work better in all lowercase. Thus, while it is convenient to use this convention to more easily distinguish when a table is in R vs. when it is in PostgreSQL, it is also important to follow precisely this convention. There are ways around this should you need uppercase table names server side. However, as this text is not primarily about SQL, it seems simpler to simply follow this lowercase convention:

```
dbWriteTable(conn = conSuper, name = "address", value = ADDRESS)
```

```
## [1] TRUE
```

```
dbWriteTable(conn = conSuper, name = "demographics", value = DEMOGRAPHICS)
```

```
## [1] TRUE
```

```
dbWriteTable(conn = conSuper, name = "thricedailymetrics", value =
THRICEDAILYMETRICS)
```

```
## [1] TRUE
```

```
dbWriteTable(conn = conSuper, name = "dailymetrics", value = DAILYMETRICS)
```

```
## [1] TRUE
```

```
rm(ADDRESS, DEMOGRAPHICS, THRICEDAILYMETRICS, DAILYMETRICS)
```

To confirm the tables we wrote to our server are indeed stored there, the function dbListTables() can be used. Notice the argument it takes is the connection (conSuper) to the entire database. Beyond only confirming that tables were uploaded, this useful function can help remind us what tables are contained in a database:

```
dbListTables(conSuper)
## [1] "address"            "demographics"         "thricedailymetrics"
## [4] "dailymetrics"
```

That is all you need to add data from R to the database. One table at a time, we can safely store information on the server.

10.5 Creating Users with Specific Privilege

Confidentiality of data is one of the three benefits we mentioned for running a database. While password-protected access is one aspect of that confidentiality, there is more than only passwords. Adding more than one user can add a helpful layer of precise information protection. Each user can be given only specific types of access to only specific tables. So in addition to removing users who no longer require any access, user's *privileges* can be fine-tuned to encourage data confidentiality.

Creating a new user requires our *super* user's privileges. Keep in mind PostgreSQL is part of the SQL language ecosystem. Thus, other than a handful of DBI-specific functions, more advanced or nuanced commands must be in SQL. The generic DBI wrapper for such SQL code is dbGetQuery(). This function takes two arguments. The first is the authenticated connection, and the second is the SQL code wrapped in quotes as a text string. The function both sends the text string to the database for processing and collects and returns any resulting output back into R.

To avoid discussing SQL in depth, we focus on sharing the format of creating a new user and giving that user specific privileges. To create a new user, the format is Create USER username WITH PASSWORD 'textinsinglequotes'. We will use the SQL convention of uppercase SQL commands and lowercase variables.

Here, we create three new grad student users to help with our research project. The first user we will only give *read* privileges (and only on one specific table). The second user will gain *insert* rights (and only on one specific table). Our third user is able to *update* and read one table. Of course, in real life, you may grant a single user more or less access to any combination of tables. Here, we will keep user permissions fairly simple:

```
dbGetQuery(conSuper, "CREATE USER gradstudentr WITH PASSWORD 's3cur3'")

## data frame with 0 columns and 0 rows

dbGetQuery(conSuper, "CREATE USER gradstudenti WITH PASSWORD 'saf3'")

## data frame with 0 columns and 0 rows

dbGetQuery(conSuper, "CREATE USER gradstudentu WITH PASSWORD 'sur3'")

## data frame with 0 columns and 0 rows
```

Once you have one or more users created, it is time to GRANT specific privileges ON specific tables TO specific users. Privilege or rights options include SELECT for read-only access and UPDATE or INSERT for two different types of write access. INSERT grants the ability to add a new row to an existing table. UPDATE grants the ability to change values in existing rows in a table. However, because one usually needs to only update certain rows, the SELECT privilege helps:

```
dbGetQuery(conSuper, "GRANT SELECT ON dailymetrics TO gradstudentr;")
```

```
## data frame with 0 columns and 0 rows
```

```
dbGetQuery(conSuper, "GRANT INSERT ON address TO gradstudenti;")
```

```
## data frame with 0 columns and 0 rows
```

```
dbGetQuery(conSuper, "GRANT SELECT, UPDATE ON address TO gradstudentu;")
```

```
## data frame with 0 columns and 0 rows
```

There is a syntax to notice about the SQL structure in the preceding code. The layout is GRANT **PRIVILIGES** ON **table** TO **username**. SQL allows commas, and the options for textbfPRIVILIGES are any combination of SELECT, UPDATE, or INSERT.

Now we have three new users whom we must connect to our database. In real life, in a single R instance, one would rarely (if ever) have more than one connection to the same database. Instead, each user normally represents a distinct person with distinct data needs/roles. Notice in the preceding two code blocks, any combination of users can be CREATEd, and any combination of privileges can be GRANTed on any combination of tables. GRANT is not a one-time event—a user may, over time, gain or lose access depending on what makes sense. In our example of a university research lab, it may be that new students have read-only access to only a few tables. Over time, as knowledge of the research develops, a student might need access to more areas of the research and even require write access as they perform their own work.

That said, best practice in user privilege is *role* based, rather than seniority. In any case, we name our two new connections conRead and conWrite. As you proceed through the remainder of this section, pay careful attention to *which* user's connection is being used:

```
conRead <- dbConnect(drv, dbname = "postgres", host = "localhost",port = 5432,
                 user = "gradstudentr", password = "s3cur3" )

conInsert <- dbConnect(drv, dbname = "postgres", host = "localhost",port = 5432,
                 user = "gradstudenti", password = "saf3" )

conUpdate <- dbConnect(drv, dbname = "postgres", host = "localhost",port = 5432,
                 user = "gradstudentu", password = "sur3" )
```

Our first order of business now these three new connections (with various levels of access) are established is to understand what each of these users can do. Much like our super user, all users can use dbListTables() to see all the tables in the database. This may seem counterintuitive, as these users were *only* given specific privileges on one table each. In the generic database structure we created, any user will be able to see the table names in the database:

```
dbListTables(conRead)
```

```
## [1] "demographics"       "thricedailymetrics" "dailymetrics"
## [4] "address"
```

```
dbListTables(conInsert)
```

```
## [1] "demographics"       "thricedailymetrics" "dailymetrics"
## [4] "address"
```

```
dbListTables(conUpdate)
```

```
## [1] "demographics"       "thricedailymetrics" "dailymetrics"
## [4] "address"
```

Despite their ability to *see* table names, our users are **not** as powerful as our super user. They cannot remove tables. While we do not demonstrate this here yet for our super user, rest assured dbRemoveTable() would work for that user. Now, we might rightly expect our gradstudentr, with read-only privileges, and on dailymetrics at that, to not be able to remove a table.

However, note well our gradstudenti and gradstudentu users, who do have UPDATE or INSERT on address, cannot remove that table. This may well be the first good hands-on example of why a database might be better than a flat table file (e.g., CSV).

While these users have some level of access, their ability to make a wholesale change (accidental or otherwise) is limited:

```
dbRemoveTable(conRead, "address")

## Error in postgresqlExecStatement(conn, statement, ...) :
## RS-DBI driver: (could not Retrieve the result : ERROR: must be owner of
   table address
## )

## Warning in postgresqlQuickSQL(conn, statement, ...): Could not create
   execute: DROP TABLE "address"

## [1] TRUE

dbRemoveTable(conInsert, "address")

## Error in postgresqlExecStatement(conn, statement, ...) :
## RS-DBI driver: (could not Retrieve the result : ERROR: must be owner of
   table address
## )

## Warning in postgresqlQuickSQL(conn, statement, ...): Could not create
   execute: DROP TABLE "address"

## [1] TRUE

dbRemoveTable(conUpdate, "address")

## Error in postgresqlExecStatement(conn, statement, ...) :
##   RS-DBI driver: (could not Retrieve the result : ERROR: must be owner
     of table address
## )

## Warning in postgresqlQuickSQL(conn, statement, ...): Could not create
   execute: DROP TABLE "address"

## [1] TRUE
```

Our users can list all database table *names* and cannot remove any tables. So far, so good. Still, those are not exactly exciting features. What about actual table access? Keep in mind each user has one or more abilities to either SELECT, UPDATE, or INSERT. It is time to learn a bit of SQL. We will discuss the specifics of these structures again later in this section. However, SQL is a very verbal language, so even if we do not yet discuss the exact meaning, it will make sense what is happening with these simple commands.

Our gradstudentr has SELECT privileges on dailymetrics. Using our R wrapper function dbGetQuery(), we send two slightly different SQL statements to our database server. The only difference in each lies in the table. Notice one works and the other does not. The reason is precisely because our conRead connection has only gradstudentr's privileges, which are only good on the table dailymetrics. The * is a wildcard character for column names. Thus, all column names are shown for allowed tables.

This command will not work. Our user does not have permission to select from the demographics table:

```
dbGetQuery(conRead, "SELECT * FROM demographics;") #does not work
```

```
## Error in postgresqlExecStatement(conn, statement, ...) :
## RS-DBI driver: (could not Retrieve the result : ERROR: permission denied
   for table demographics
## )
```

```
## Warning in postgresqlQuickSQL(conn, statement, ...): Could not create
   execute: SELECT * FROM demographics;
## NULL
```

This command will work. Our user does have permission to select from the dailymetrics table. To save space in the text, we assign the results to the variable query and then show only the first few rows:

```
query <- dbGetQuery(conRead, "SELECT * FROM dailymetrics;")
head(query)
```

##	row.names	userid	surveyday	sols	wasons	support	copeprb	copeprc
## 1	1	1	2017-02-24	NA	NA	NA	NA	NA
## 2	2	1	2017-02-24	0.0	0	7.0	2.3	2.4
## 3	3	1	2017-02-25	NA	NA	NA	NA	NA
## 4	4	1	2017-02-25	NA	NA	NA	NA	NA

## 5	5	1 2017-02-25	6.9	0	6.2	NA	NA
## 6	6	1 2017-02-26	NA	NA	NA	NA	NA

##	copeexp	copedis
## 1	NA	NA
## 2	2.4	2.2
## 3	NA	NA
## 4	NA	NA
## 5	2.0	NA
## 6	NA	NA

Our gradstudenti has only INSERT privileges on address; our gradstudentu has both SELECT and UPDATE privileges on address.

Using our R wrapper function dbGetQuery(), we send two identical SQL statements to our database server. The only difference in each lies in the user. Notice one works and the other does not. The reason is precisely because our conInsert connection has gradstudenti's privileges, which are only good for adding a row or *inserting* on the table address. This user does not have SELECT rights.

While these different user privileges are slightly made up in this case, they showcase various levels of access. By making use of the different combinations of privileges, you most likely can find a workable level of security and access for your needs.

This command will not work. Our user only has permission to INSERT (e.g., bind new rows to the end of the table) to address:

```
dbGetQuery(conInsert, "SELECT * FROM address;")
```

```
## Error in postgresqlExecStatement(conn, statement, ...) :
## RS-DBI driver: (could not Retrieve the result : ERROR: permission denied
   for table address
## )
```

```
## Warning in postgresqlQuickSQL(conn, statement, ...): Could not create
   execute: SELECT * FROM address;
```

```
## NULL
```

This command will work. gradstudentu has permission to both SELECT and UPDATE from the address table. To save space in the text, we assign the results to the variable query and then show only the last few rows:

```
query <- dbGetQuery(conUpdate, "SELECT * FROM address;")
tail(query)
```

```
##       row.names userid postalcode
## 186         186    186       3807
## 187         187    187       4118
## 188         188    188       3843
## 189         189    189       3952
## 190         190    190       4181
## 191         191    191       4582
```

So what *can* gradstudenti do? They can INSERT into address. The command is INSERT INTO table VALUES (list, values, here, separated, by, commas, one, value, for, each column). Because our gradstudenti user cannot select and see what is in the table, we use our super user to view the updated table values:

```
dbGetQuery(conInsert, "INSERT INTO address VALUES (200, 200, 9999)")
```

```
## data frame with 0 columns and 0 rows
```

```
query <- dbGetQuery(conSuper, "SELECT * FROM address")
tail(query)
```

```
##       row.names userid postalcode
## 187         187    187       4118
## 188         188    188       3843
## 189         189    189       3952
## 190         190    190       4181
## 191         191    191       4582
## 192         200    200       9999
```

Notice our table now has 192 rows instead of only 191 rows.

So what *can* gradstudentu write to a table? They can SELECT or UPDATE into address. Keep in mind the address table has columns for userid and postalcode. These userids are for our study participants. Suppose the first participant changed their address. For ease of visualizing, suppose their postal code changed to 1111.

Here, we add an extra phrase to our SQL sentence: WHERE userid = 1. This allows us to only display the row(s) about the first participant. While it is a useful space-saving device for our SELECT statement, it is vital for our UPDATE statement (otherwise, **all** rows would be updated to a postalcode of 1111):

```
dbGetQuery(conUpdate, "SELECT * FROM address WHERE userid = 1")
```

```
##   row.names userid postalcode
## 1         1      1       4112
```

```
dbGetQuery(conUpdate, "UPDATE address SET postalcode = 1111 WHERE userid = 1")
```

```
## data frame with 0 columns and 0 rows
```

```
dbGetQuery(conUpdate, "SELECT * FROM address WHERE userid = 1")
```

```
##   row.names userid postalcode
## 1         1      1       1111
```

Having shown some nuances of users with various permissions, we conclude this section by demonstrating how to revoke access from users. Prior to deleting or DROPping a user entirely, one must revoke all their access. Notice, however, that the REVOKE statement, much like GRANT, can be customized. In other words, a user's privileges can be modified regularly, depending on what makes sense for any given timeframe.

First, we disconnect our three users' connections using dbDisconnect() and clear those variables from our global environment:

```
dbDisconnect(conRead)
```

```
## [1] TRUE
```

```
dbDisconnect(conInsert)
```

```
## [1] TRUE
```

```
dbDisconnect(conUpdate)
```

```
## [1] TRUE
```

```
rm(conRead, conInsert, conUpdate)
```

Notice we leave our super user in place. This is required to make our final changes to the database. While, in this case, we do specifically REVOKE each privilege granted to each user on each table (again to demonstrate privileges can be more dynamic), we could use ALL in place of a list of privileges:

```
dbGetQuery(conSuper, "REVOKE SELECT ON dailymetrics FROM gradstudentr;")
```

```
## data frame with 0 columns and 0 rows
```

```
dbGetQuery(conSuper, "REVOKE INSERT ON address FROM gradstudenti;")
```

```
## data frame with 0 columns and 0 rows
```

```
dbGetQuery(conSuper, "REVOKE SELECT, UPDATE ON address FROM gradstudentu;")
```

```
## data frame with 0 columns and 0 rows
```

Now that our users have no more rights, we can DROP USER our users from the database. After this, they are gone and cannot connect to the server anymore. Their records have been removed:

```
dbGetQuery(conSuper, "DROP USER gradstudentr")
```

```
## data frame with 0 columns and 0 rows
```

```
dbGetQuery(conSuper, "DROP USER gradstudenti")
```

```
## data frame with 0 columns and 0 rows
```

```
dbGetQuery(conSuper, "DROP USER gradstudentu")
```

```
## data frame with 0 columns and 0 rows
```

To demonstrate this is the case, we show this example of what one of our users might see should they try to reconnect to the server after being DROPped:

```
dbConnect(drv, dbname = "postgres", host = "localhost",port = 5432,
                user = "gradstudentr", password = "s3cur3" )
```

```
## Error in postgresqlNewConnection(drv, ...): RS-DBI driver: (could not
   connect gradstudentr@localhost:
password authentication failed for user "gradstudentr"
## )
```

As you see, a database can have multiple users with different levels of access to any collection of tables in the database. The administrator or super user can add and remove users as needed, as well as change their access rights.

10.6 Password Security

It is worth noting that in our prior examples of connecting to a database, we engaged in the rather unsafe practice of writing our password in *plain text*. Even worse, far from sourcing it from a file on our computer, it is written in our code itself. Simply sharing this code would leak our password.

Most operating systems, including Windows, Mac OS, and Unix, have a more secure area to store usernames and passwords. Digital security changes swiftly, and there may well be other, easier, or more secure ways to store passwords and authenticate access to a database. That said, we present one possible solution that at the very least can prevent a plain-text password hard-written in your code.

The R package keyring has a set of functions which can store a password for a particular service and user in a more secured area of your operating system. It is important to note this password would then not be available on a different operating system on another machine.

For the sake of demonstration, we will disconnect our super user, store that password in the key store, and show how to connect using the key store to add a layer of safety to your database access:

```
dbDisconnect(conSuper)
```

```
## [1] TRUE
```

```
rm(conSuper)
```

The function key_set_with_value takes four arguments. The service can be used to organize different users for different services. In our case, we call this the advancedrBookSQLch10 database. The username will be how a particular user's password is retrieved from the key store. In this case, we use our super user's name, postgres. Lastly, we use our super user's password, advancedr.

It is worth noting that while this creates the initial record, running the same code with a different password would simply update the password:

```
key_set_with_value(service = "advancedrBookSQLch10",
                   username = "postgres",
                   password = "advancedr")
```

Having set the password, we can confirm that a new key store record was created by using the `key_list()` function. Notice our function does not show the password—it simply shows a record was stored:

```
key_list()
```

```
##                    service username
## 1 advancedrBookSQLch10 postgres
```

Notice that what the preceding code confirms is that our operating system is remembering our password for us now. There is, in general, no need to run the prior two blocks of code more than once. The only time it would be required to use that code would be to either add your password to a new local system or to update a password.

More often, to get your password for use, it is saved into a local variable, such as usrPassword via the `key_get()` function. The `service` and `username` must match one on record. Now, it is important to understand the password is stored in your global R environment as plain text. Someone looking over your shoulder could still see your password! The following two blocks of code are what is most often run once per session to connect to our database:

```
usrPassword <- key_get(service = "advancedrBookSQLch10",
                       username = "postgres")
```

With your password now available to R, we run our connection protocol again. The only difference is that this time, rather than the plain-text password in quotes, we use our variable name `password = usrPassword`. Once the connection is established, we delete the password from the global environment. This is not a secure delete mind; it simply prevents onlookers from visually seeing the password:

```
conSuper <- dbConnect(drv, dbname = "postgres", host = "localhost",port = 5432,
                      user = "postgres",
                      password = usrPassword )
rm(usrPassword)
```

The connection is reestablished, and we are ready to do whatever work might be required. Just in case you ever needed to delete the password and record from your operating system, we show the following code. Again, much like adding a key, this is a once-per-system bit of code:

```
key_delete("advancedrBookSQLch10", "postgres")
key_list()
```

```
## [1] service username
## <0 rows> (or 0-length row.names)
```

10.7 SQL from R

The DBI package has some functions that allow us to achieve results without directly using SQL. We already met the dbListTables() function which shows all tables in a database. The dbListFields is similar to R's colnames():

```
dbListTables(conSuper)
```

```
## [1] "demographics"      "thricedailymetrics" "dailymetrics"
## [4] "address"
```

```
dbListFields(conSuper, "thricedailymetrics")
```

```
## [1] "row.names"          "userid"             "surveyinteger"
## [4] "surveystarttimec11" "stress"             "posaff"
## [7] "negaff"
```

A nice feature of these functions is they do what they say. While earlier we showed how to obtain the output of a table using pure SQL, there is also a function for reading a table, dbReadTable(). Deciding how much SQL to use, and what operations to perform server side vs. RAM side in R, is a choice. If this is your first time using a second programming language, you will likely take slow steps moving between these two worlds. If you use an SQL database regularly, as one of your authors had occasion to start a few years back, you may find that over time, certain operations make more sense to complete server side.

In most scenarios, there is no significant difference between dbGetQuery(conSuper, SELECT * FROM thricedailymetrics) and our following function:

```
query <- dbReadTable(conSuper, "thricedailymetrics")
head(query)
```

```
##    userid surveyinteger surveystarttimec11 stress posaff negaff
## 1      1             2             1.9e-01      5    1.5    1.7
## 2      1             3             4.9e-01      1    1.5    1.0
## 3      1             1             1.2e-05      1    1.6     NA
## 4      1             2             1.9e-01      2    1.6    1.4
## 5      1             3             4.1e-01      0    1.1    1.0
## 6      1             1             1.6e-02      0    1.3    1.7
```

Lastly, while we already wrote several tables to our database, we draw your attention again to the dbWriteTable() function. This function takes both an overwrite and an append option set to FALSE or TRUE. This is similar to UPDATE vs. INSERT. Both are convenient in the right situation. We first write the iris dataset to our table, showing the *default* options for both these variables. Notice if we run our function a second time, it returns an error:

```
dbWriteTable(conn = conSuper, name = "iris", value = iris, overwrite =
FALSE, append = FALSE)
```

```
## [1] TRUE
```

```
dbWriteTable(conn = conSuper, name = "iris", value = iris, overwrite =
FALSE, append = FALSE)
```

```
## Warning in postgresqlWriteTable(conn, name, value, ...): table iris
exists in database: aborting assignTable
```

```
## [1] FALSE
```

Should you wish to append (e.g., INSERT) data at the end of the frame, simply set append = TRUE. On the other hand, if you need to fully overwrite (e.g., UPDATE all elements), then instead set overwrite = TRUE.

Recall the basic SQL query SELECTs specific columns, FROM a specific table. We use dbGetQuery() to send our query to and execute it in PostgreSQL. In this way, a table from our database can be brought into R:

```
query <- dbGetQuery(conSuper,
                    statement = "SELECT *
                                 FROM address")
query <- as.data.table(query)
head(query)
```

```
##      row.names userid postalcode
## 1:           2      2       4132
## 2:           3      3       3880
## 3:           4      4       3781
## 4:           5      5       4058
## 5:           6      6       4627
## 6:           7      7       3996
```

Remember, the * was a wildcard. If we needed only one column of data, we would SELECT that specific column rather than using a wildcard value:

```
query <- dbGetQuery(conSuper,
                    "SELECT postalcode
                     FROM address")
query <- as.data.table(query)
head(query)
```

```
##      postalcode
## 1:         4132
## 2:         3880
## 3:         3781
## 4:         4058
## 5:         4627
## 6:         3996
```

Recall we had (for the UPDATE command at the time) used the WHERE on a column value. Keep in mind this could just as easily be done in R using query[userid = 1]. However, one reason to perhaps use this in SQL rather than in R might be size in memory. Some tables could be very large, and by selecting only certain rows, the size in RAM can be reduced:

```
query <- dbGetQuery(conSuper,
                    "SELECT postalcode
                     FROM address
                     WHERE userid = 1")
query <- as.data.table(query)
head(query)

##      postalcode
## 1:         1111
```

As we saw when assigning privileges, more than one item can be done at a time in SQL using a comma. In this case, from the `thricedailymetrics` table, we `SELECT` two columns with a more advanced `WHERE` clause:

```
query <- dbGetQuery(conSuper,
                    "SELECT userid, stress
                     FROM thricedailymetrics
                     WHERE surveyinteger >= 2")
query <- as.data.table(query)
head(query)

##      userid stress
## 1:        1      5
## 2:        1      1
## 3:        1      2
## 4:        1      0
## 5:        1      3
## 6:        1      1
```

One purpose perhaps of PostgreSQL is to allow for data larger than RAM might physically allow. It can be read in chunks as we demonstrated, or we can push large amounts of data from R to free up memory. Our next example comes with a choice. If you have enough available free space in your system RAM and hard disk, go ahead and do the example of 200 million rows as written. Otherwise, change the number of variables in `rnorm()` to something lower, perhaps a smaller example of 70 million rows.

This code creates 200 million entirely made-up data observations with a mean of 1 and a default standard deviation of 1. As you see, it is about 1.6 GB in size (hover over the value in the global environment pane in RStudio). Certainly, data can be larger, but this might be good enough as a test case to see how it works on your system. It likely takes a few minutes to run, so be prepared for a bit of a wait:

```
BiggerData <- rnorm(200000000, 1)
BiggerData <- as.data.table(BiggerData)
dbWriteTable(conSuper, "biggerdata", BiggerData)

## [1] TRUE

rm(BiggerData)
dbListTables(conSuper)

## [1] "demographics"      "thricedailymetrics" "dailymetrics"
## [4] "address"           "iris"               "biggerdata"
```

Again, this may have taken a handful of minutes to run on your system. However, if you succeeded, you have a sense of how much data can be transferred to a database. As we have no use for this BiggerData (other than a "stress test" of our system), we use the function dbRemoveTable():

```
dbRemoveTable(conSuper, "biggerdata")

## [1] TRUE

dbListTables(conSuper)

## [1] "demographics"      "thricedailymetrics" "dailymetrics"
## [4] "address"           "iris"
```

While we are (at least with super user privileges) able to delete a table, it does not necessarily release space to us in PostgreSQL (and hence on our hard drive). Now, with all the extra space available, new writes into our database would not take up more disk space. However, if we need to force a cleanup of "dead" bytes, SQL has a VACUUM FULL operation which shrinks the database down to only active data. In our case, this would not take all that long to achieve. Ours is a simple structure. More complex databases and servers can find this operation somewhat costly. Thus, this is a database maintenance operation that may only occur periodically or as needed:

```
dbGetQuery(conSuper, "VACUUM FULL")
```

```
## data frame with 0 columns and 0 rows
```

With that, we have learned quite a bit about using an SQL database with R. While SQL is a language in its own right, this ought to be enough to get started on doing data analysis and data maintenance for advanced users of R.

We include this second-to-last block of code only for completeness. By deleting these tables from your SQL database, you are now able to run all prior code again and get the same results. Otherwise, attempting to write a table (as we did to populate our database early on in the chapter) would cause an error:

```
dbRemoveTable(conSuper, "address")
```

```
## [1] TRUE
```

```
dbRemoveTable(conSuper, "demographics")
```

```
## [1] TRUE
```

```
dbRemoveTable(conSuper, "thricedailymetrics")
```

```
## [1] TRUE
```

```
dbRemoveTable(conSuper, "dailymetrics")
```

```
## [1] TRUE
```

```
dbRemoveTable(conSuper, "iris")
```

```
## [1] TRUE
```

```
dbGetQuery(conSuper, "VACUUM FULL")
```

```
## data frame with 0 columns and 0 rows
```

To close this connection, there are two steps rather than one, because we must disconnect from PostgreSQL and unload our driver that allows DBI to make that connection. If successful, they will both return TRUE:

```
dbDisconnect(conSuper)
```

```
## [1] TRUE
```

```
dbUnloadDriver(drv)
```

```
## [1] TRUE
```

```
rm(conSuper, drv, query)
```

This concludes our demonstration of PostgreSQL. The real way to make the knowledge you now have more powerful is to delve further into SQL syntax to use dbGetQuery() on ever more exotic quests. For now, though, a small-enough quantity of data ought to be readable into R that some analytics can happen.

10.8 Summary

This chapter gave you direct access to SQL database with R. While many of the features are not designed for curating data, R nevertheless provides a quick way to get a handle on big data. Such data sets seem to be becoming ever more common, and while it may not be feasible for researchers to be familiar with every database, that is not required. Using R, it is entirely possible to pull out a subset of data that contains all items of interest and is small enough to fit into working memory. Skilled users of R, such as this book's readers, can prune the data in R itself further and from there perform desired analytics. Table 10-1 provides a summary of the key functions and phrases in this chapter.

Table 10-1. *Listing of key functions described in this chapter and summary of what they do*

Function	What It Does
dbConnect()	Uses the DBI package plus a driver to connect to a PostgreSQL database.
dbDriver()	Driver needed for PostgreSQL (and others).
dbWriteTable()	Writes data from R to PostgreSQL (you must be a user with write privileges).
dbListTables()	Lists all tables in a database.
dbListFields()	Lists all fields (column names) in a database.
dbGetQuery()	Sends an SQL query to a PostgreSQL database and executes the query there.
dbRemoveTable()	Drops a table from PostgreSQL (user must have privileges); does not necessarily VACUUM.
dbDisconnect()	Closes the DBI package connection to PostgreSQL.
dbUnloadDriver()	Unloads the PostgreSQL drivers.
key_set_with_value()	Uses the keyring package to store a password.
key_list()	Lists keys stored in the operating system's key store.
key_get()	Retrieves a specified key from the OS key store.
CREATE	SQL to create a user.
DROP	SQL to drop a user (e.g., delete).
GRANT	SQL to grant a user certain privileges.
REVOKE	SQL to revoke certain privileges from a user.
SELECT	SQL to select data from a table.
UPDATE	SQL to update data inside a table.
INSERT	SQL to append data to a table.

CHAPTER 11

Getting Your Cloud

Depending on your needs and uses for R, it can be convenient to have a reasonable amount of memory and processor capability. Often this power is necessary only on occasion, and it may not be cost-effective to own the hardware. This is where hosting R on the cloud may be helpful. Cloud instances bring on-demand resources that are readily scaled up or down as each situation requires. These days, there are several groups that provide such services at very reasonable prices. The challenge we face is threefold, and we walk through those steps over the next three chapters. We need to get a cloud, we need to administer our cloud's operating system (in some cases), and we need to do some fun things with R.

This chapter focuses on getting your own cloud. To do that, we need to choose a provider, start up *instances* (hardware/software) in a location we never *physically* access, and successfully connect to our instances. Since the publication of the first edition of this book, the process has become increasingly user-friendly.

This guide is somewhat based on the idea that your local system is a Windows-based computer. We assume you are most familiar with a Windows environment. We do not suppose you have administrative privileges on your local machine, although it requires a network connection. You also need access to a credit card. Although your work in this chapter is meant to use (as of the time of this writing) on free-tier services, that may change based on options selected or provider decisions. In other words, when we wrote this, our intention is this stage would not involve costs for learning-level usage. Finally, and apologies for those for whom this is not true, we are going to assume that you have no prior knowledge of clouds, networks, or Unix-flavored operating systems. Thus, we will work our way fairly slowly through each step.

Most steps here should also work from a local Apple machine. The steps that may not work are not required to set up cloud access. Thus, if your main local device is from Apple, you can definitely still put R on the cloud.

© Matt Wiley and Joshua F. Wiley 2020
M. Wiley and J. F. Wiley, *Advanced R 4 Data Programming and the Cloud*,
https://doi.org/10.1007/978-1-4842-5973-3_11

11.1 Disclaimers

There are a few disclaimers we should make at this point. We are in no way endorsing Amazon Web Services (AWS) [3] just as we are in no way endorsing Windows. Nevertheless, AWS is convenient, popular, and, as of this writing, offers a free tier of service for 12 months (`http://aws.amazon.com/free/`). Please be careful with data you place in the cloud; without proper attention to security, it might be easy for that information to find its way to the wrong minds. While we give some brief suggestions on digital safety, these are not enough. The topic of Internet security can and does fill several books; if you have data that should not be faxed or emailed, there is much more to be learned before processing that data on a cloud. And again, while it is not our intention to write steps that would require costs, we do not control the AWS business model.

With luck, we have not scared you away from this convenient and powerful method of bringing just the right resources to your research and data. We also recommend reading this chapter and the one following first and dedicating an afternoon or more to the process if you are new to the cloud. We should also mention many of this chapter's screenshots may look slightly different on your system (in part due to browser choice). Indeed, AWS itself has semi-regular layout changes as new features are added or old functionalities merge. Nevertheless, what you see should look quite close to what we show.

11.2 Starting Amazon Web Services

Unfortunately, just because one is comfortable with R does not ensure that moving to the cloud is instant. All the same, the system is fairly well designed to start up quickly after some legwork is done. Our goal is to help you access an Elastic Compute Cloud (EC2) instance. EC2 is a virtual computer that can be customized to have whatever processing power and memory combination you may wish. That is an overly bold statement. However, at the time of this writing, anywhere from 1 processor at 2.5 GHz and 0.5 GiB of memory to 128 processors at 2.3 GHz and 3904 GiB of memory can constitute a single computer instance. Of course, for the free tier, the latter is not an option.

The first step we take is visiting the Amazon Web Services site (`http://aws.amazon.com/free/`) and getting a new account, as shown in Figure 11-1. Follow the steps to create a new account, being sure to generate a secure password. This step requires both a credit card and a phone number. Again, our choice of browser vs. your browser's view may create visual differences on occasion.

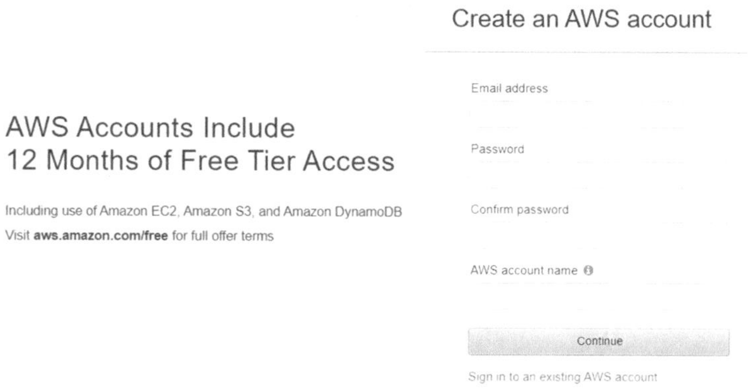

Figure 11-1. *Creating a new AWS account*

The account you just created is your top-level account. The recommendation is to create a sub-account that does not have the full rights this one does. In fact, Amazon goes to some pains to convince you to use *Identity and Access Management* (IAM) to create a secondary account. If you are just a single user who never needs someone else to access this, a secondary account is still beneficial. If you are ever going to have other users, definitely create this second, non-root account for yourself. From adding a graduate student to adding a colleague to adding employees or consultants, it can be helpful to not give away the "master key" to the "castle." Do not be shocked by the plethora of options available; simply scroll down to the *Security, Identity, & Compliance* portion of AWS and select *IAM* from near the end of the third column, as shown in Figure 11-2. Depending on your browser, *IAM* may have a key icon or may have no icon at all. There are enough options present that pressing *CTRL+F* and then typing *IAM* into the search box may help highlight the text for IAM. Do not be shy about searching, clicking, and exploring to find it. You cannot hurt anything, and Amazon updates the interface fairly regularly.

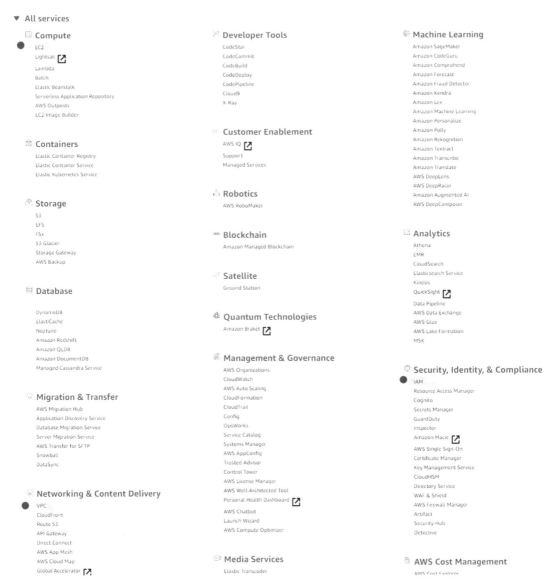

Figure 11-2. *The AWS management console with highlighted boxes and dots for the three parts we use in this chapter*

Inside IAM, you need to create both a new group and your first user. *Groups* are convenient because they allow you to create multiple users who have only certain privileges. On the left side of the screen, select the *Groups* option and then *Create New Group* at the top of the loaded screen. We called ours *AdvancedR*, although yours might be more sensibly named CloudAdmin or something similar. When asked which policy, we recommend *AmazonEC2FullAccess*, as shown in Figure 11-3.

Figure 11-3. *IAM group policy selection—your view may differ*

Next, we create a user to live in our group. Keep in mind, as a user in the "full access" group, this account will have a great deal of control. While following directions on the establishment of the user, we recommend using your first name so you see that account as you and using a different password from your AWS root account. Start by selecting the *Users* tab on the left side. Next, select *Add user* from the top of the screen that loads. Once there, create your username. While you can create more than one user at a time (using *Add another user*), it may be best to save that for later—especially if this is your first time on AWS. Be sure the settings are as in Figure 11-4, including enabling console access. You could choose to create a custom password directly, of course.

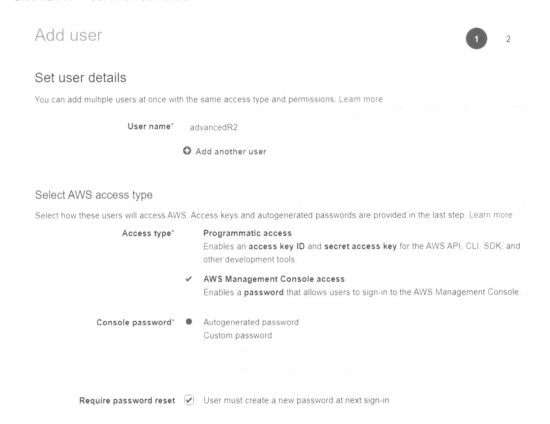

Figure 11-4. Creating a user on the AWS IAM

When you created your account initially, AWS gave you an account identification number. To see this number and link again, click the IAM *Dashboard* at the top left. The dashboard has an *IAM users sign-in link* as well as an option to customize that to a human-friendly name if you wish. Now, sign out of AWS. Go to your AWS address that follows this pattern: `https://YOUR_ID_HERE.signin.aws.amazon.com/console`. Sign in using your new username and password. You may have noticed that inside IAM, there were also options to add multifactor security to your root account. That is most likely a wise move. By ensuring your root account is secure, even if a lower privileged account were compromised or "hacked," you would at least be able to clean up the damage afterward. The steps are easy enough to follow, although again beyond the scope of this rapid setup discussion.

11.3 Launching an EC2 Instance

Taking a look at Figure 11-2 again, notice in the top-left corner the link for *EC2* under *Compute*. Click this link; at the top right, you can select a region such as Oregon or Sydney. Different regions tend to have variations in cost, and of course, the closer you are geographically to a region, the more likely network upload and download speeds are efficient. As is often the case, there are trade-offs to consider; we recommend selecting the region nearest to you. One warning is that after you select a region, some of what we do is regionally specific. In particular, key pairs (used to access your cloud server) tie to particular regions.

Once you have your region selected, it is time to create a key pair. On the left side is a menu list, and under *Network and Security* is the *Key Pairs* link. Select that link, and then click the *Create key pair* button. Give your key pair a name. If you are likely to need cloud instances in different regions, Amazon's advice to add your region name to the key pair name is sound. Go ahead and choose the *ppk* version for use with PuTTY. Then, one last time, select *Create key pair*. Your browser should automatically download the key pair file. Remember where this is saved and keep it safe; this is your only chance to download this file. We use it later in the next section, but for now, we move on to creating a virtual private cloud (VPC). Return to the home screen of Figure 11-2.

One last look at Figure 11-2 shows that in the first column, near the end, under *Networking*, is *VPC*. There are several options here; we go with the simplest, one with a single public subnet. Select the *Launch VPC Wizard* and give your VPC a name such as AdvancedR. Stick with the default options and then select *Create VPC*. As you can see from these options, if you were interested in creating a more advanced structure, that is readily done.

Your cloud is not safe without a security group. We are very cautious with our security group. On the left in the *VPC Dashboard*, notice the *Security* tab and the *Security Groups* link. Click that link and then click the *Create Security Group* button. Give the group a name tag and a name such as AdvancedR, as well as a description. Be sure to select your VPC from that drop-down list at the end and then click the *Create* button.

Now, select your AdvancedR security group and notice the options that are at the lower end of your browser screen, shown in Figure 11-5. What we are about to do is configure which IP addresses are allowed to visit your future cloud instance. While not an ironclad layer of security, it can help. Certainly, this is a good first step, as it makes it more complex for people who are not you to know your server even exists. Be sure to select the *Inbound Rules* tab, select *Edit rules*, and add SSH (22) with TCP protocol; you want to put your IP address for the source. For now, go ahead and use *My IP*. If you later need to add another address, you can. For example, to write this book, both authors' IP addresses were added. It should be in valid Classless Inter-Domain Routing (CIDR) notation, which means that if your local computer's address is 72.18.154.27, you want to type in 72.18.154.27/32 for *Source*. While we are here, we also create rules for ports 80, 443, 3838, 5432, and 8787—which are HTTP, HTTPS, Shiny, PostgreSQL, and RStudio Server, respectively. However, we also lock these to just your IP address. This likely prevents others from accessing your server until you are ready. Be sure to save rules before you leave this screen. Go ahead and leave the *Outbound Rules* to allow all traffic to destination 0.0.0.0/0 (which is the world—in other words, your machine can connect *to* anywhere).

Inbound rules Info

Type Info		Protocol Info	Port range Info	Source Info		Description - optional Info
HTTP	▼	TCP	80	Cust... ▼	Q	HTTP (local computer IP only)
					72.18.154.27/32 ✕	
PostgreSQL	▼	TCP	5432	Cust... ▼	Q	PostgresSQL (local computer IP only)
					72.18.154.27/32 ✕	
SSH	▼	TCP	22	Cust... ▼	Q	SSH (local computer IP only)
					72.18.154.27/32 ✕	
Custom TCP	▼	TCP	8787	Cust... ▼	Q	Rstudio Sever (local computer IP only)
					72.18.154.27/32 ✕	
HTTPS	▼	TCP	443	Cust... ▼	Q	HTTPS (local computer IP only)
					72.18.154.27/32 ✕	
Custom TCP	▼	TCP	3838	Cust... ▼	Q	Shiny Server
					72.18.154.27/32 ✕	

Figure 11-5. *Security group's inbound rules with CIDR notation*

If your IP address changes often, you may need to relax your rules. We recommend doing so as cautiously as possible and only after you are through with the basic operating system and safety updates we discuss in the next chapter. Also notice that as of now, if you were to host a web server on your cloud instance, that website would be visible only from your local computer or network. Later, after we have installed a site, we'll go back to this screen and make some edits to allow certain parts of our server to be visible to a wider audience.

While a full lesson on CIDR is beyond the scope of this book, a little knowledge makes sense here. In "real life," the Internet does not use word addresses such as joshuawiley.com. Rather, it uses a series of 32-bit numbers. Computers use binary numbers rather than base 10; they see our website as 01001000.00010010.10011010.000 11011. That unique address connects us to a particular machine (and happens to be 32 characters long). The /32 tells AWS that only that machine is allowed; all 32 bits are fixed. Contrastingly, for outbound rules, we need our instance to be able to browse the Web at will. Thus, we use /0 to let it know that none of the bits is fixed. If you are in a corporate or university environment and want some co-workers to be able to access your instance, talk to your IT department to see what the size of your address space is. If we found our IP addresses changed frequently, we might try 72.18.154.0/24, which would allow the range of addresses 72.18.154.0–72.18.154.255 (and would nicely include the .27).

Another option is to add a handful of "duplicate" inbound rules. In other words, if you routinely access the Web from both house and work, then you could add two versions of the SSH port, one at each IP address.

We turn our attention now to creating an EC2 instance. Select the *VPC Dashboard* link at the top left, and then click the *Launch EC2 Instances* button. Select the *Ubuntu server* that is free tier eligible, as shown in Figure 11-6. You most likely want the highest version number for Ubuntu possible. In this book, we are using 18.04.

Figure 11-6. *Ubuntu server selection for starting your first EC2 instance*

Next, select the *t2.micro* type. On the *Next: configure instance details* screen, be sure to select your VPC as the *Network* setting, and set the *Auto-assign Public IP* to *Enable*. While you always have the option to *Review and Launch*, we continue through the wizard for now by selecting *Next*. As of this writing, up to 30 GiB are free. Go ahead and give yourself 30 GiB. For tags, a good name is AdvancedR again (admittedly, the name is getting a shade overused, yet it continues to work). Also, for the *security group*, be sure to *select an existing security group* as well as your AdvancedR group before finally clicking *Review and Launch*. From there, *Launch*!

Your last step is to select your key pair. Remember, we already downloaded this earlier. Be sure you have that file, and be sure to have the settings of *choosing that existing key pair* and selecting your key pair (remember, we called ours AdvancedR). Next, select the acknowledgement check box. Finally, click the *Launch Instances* button, as in Figure 11-7.

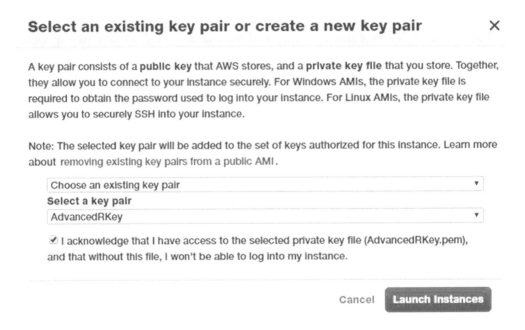

Select an existing key pair or create a new key pair   ✕

A key pair consists of a **public key** that AWS stores, and a **private key file** that you store. Together, they allow you to connect to your instance securely. For Windows AMIs, the private key file is required to obtain the password used to log into your instance. For Linux AMIs, the private key file allows you to securely SSH into your instance.

Note: The selected key pair will be added to the set of keys authorized for this instance. Learn more about removing existing key pairs from a public AMI.

Choose an existing key pair   ▾
Select a key pair
AdvancedRKey   ▾

☑ I acknowledge that I have access to the selected private key file (AdvancedRKey.pem), and that without this file, I won't be able to log into my instance.

Cancel   **Launch Instances**

Figure 11-7. *Instance key pair settings and final launch*

From here you may view your instances, which take you to the *EC2* Dashboard. It does take some time for the instance to spin up, so while we are waiting, we'll go ahead and meet you in the next section.

11.4 Accessing Your Instance's Command Line

You have an instance on AWS, and it is starting and waiting for you to connect to it. Before that can happen, there are some steps to take on Windows.

You need to download *PuTTY* [22]. You want to download the installer for Windows from `www.chiark.greenend.org.uk/~sgtatham/putty/latest.html`. Once that is done, run the installer. If your local machine does not grant you the privileges to install, you may download each of these files one at a time in the *Alternative binary files* tab. As the installer progressed, we chose to have a link to PuTTY installed on our desktop.

Using PuTTY, we may now connect to our instance. Before opening PuTTY, go back to the *EC2 Dashboard* on AWS. By now, the instance should be running, and there is some information we need from Figure 11-8. The *Public IP address* is necessary—and of course the fact that it is running. Be sure to copy the address for your instance.

Figure 11-8. *EC2 Dashboard view of our instance with Public IP address and the Actions drop-down menu*

Now we finally connect to our server for the first time. Go ahead and start PuTTY. On the navigation menu on the left, choose the SSH option and then the Auth option, and browse for your **.ppk* key, as pictured in Figure 11-9.

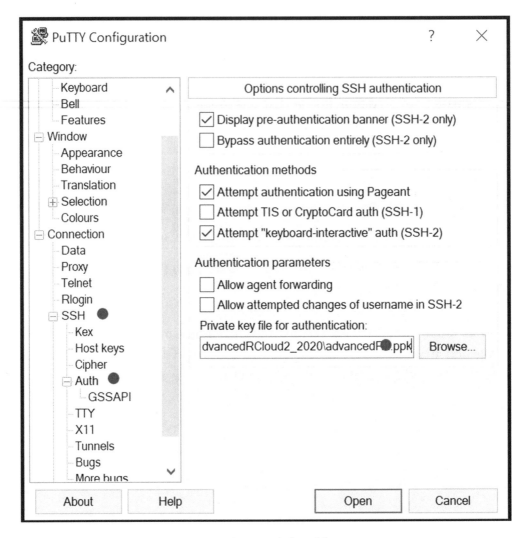

Figure 11-9. *PuTTY configuration for *.ppk key file*

Go back to *Session* in the navigation menu on the left of PuTTY, and enter ubuntu@54.201.117.64 (as in our example) or more generally *ubuntu@YOUR_EC2_ PublicIP_HERE*. Make sure the *Port* is set to 22 and the *Connection* type is *SSH*. We recommend saving your session by writing AdvancedR into the *Saved Sessions* text and clicking *Save* before proceeding. This makes it easy to access your *t2.micro* instance in the future. Go ahead and open the connection, and PuTTY gives you a security alert, as in Figure 11-10. While there is a way to compare this fingerprint with the log file of your AWS instance, the process is more than a bit involved. Confirm your acceptance by selecting *Yes*.

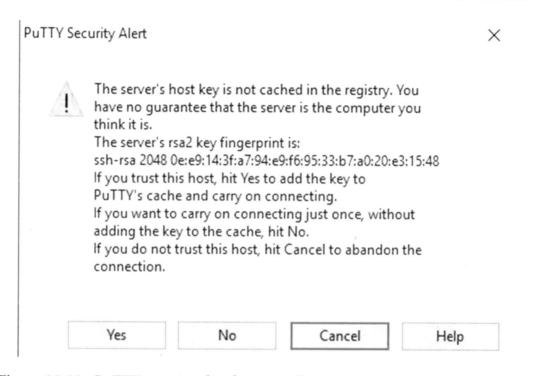

Figure 11-10. *PuTTY security alert for server fingerprint*

You are now connected to your cloud instance, and you are in the Ubuntu command line! As shown in Figure 11-11, you have access to the command line and are now ready to move on to using your server as shown in the next chapter if you wish.

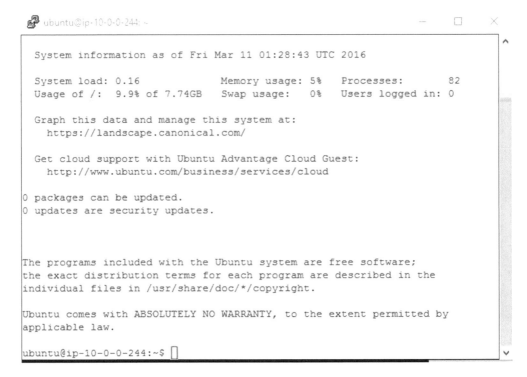

```
ubuntu@ip-10-0-0-244: ~                                      —    □    ×

  System information as of Fri Mar 11 01:28:43 UTC 2016

  System load: 0.16            Memory usage: 5%    Processes:          82
  Usage of /:  9.9% of 7.74GB  Swap usage:   0%    Users logged in: 0

  Graph this data and manage this system at:
    https://landscape.canonical.com/

  Get cloud support with Ubuntu Advantage Cloud Guest:
    http://www.ubuntu.com/business/services/cloud

0 packages can be updated.
0 updates are security updates.



The programs included with the Ubuntu system are free software;
the exact distribution terms for each program are described in the
individual files in /usr/share/doc/*/copyright.

Ubuntu comes with ABSOLUTELY NO WARRANTY, to the extent permitted by
applicable law.

ubuntu@ip-10-0-0-244:~$ []
```

Figure 11-11. *Seeing the command line through PuTTY's eyes*

We, however, wait just a bit before we get to using our server (we will walk you through this in the next chapter).

While we need to take some more steps to get R functional on the cloud, we take on one more section here to allow ourselves the luxury of uploading files to our cloud. There is one command we should teach you before we upload files. It is the `exit` command, and it is simply the word exit followed by Enter/Return. This closes your PuTTY session:

```
$ exit
```

11.5 Uploading Files to Your Instance

We use WinSCP [14] to upload files to the cloud. This program may be downloaded from `https://winscp.net/eng/download.php` (we used the Portable Executables link). If you already set up PuTTY, you may say *Yes* to the option to import sessions from PuTTY. If you have not already set up PuTTY to connect to your Ubuntu instance, that is okay. After starting WinSCP, open the initial window, which we filled in as shown in Figure 11-12. Specifically, we entered our IP address for the hostname, ensured we were set up to connect to port 22, and entered *ubuntu* as the username.

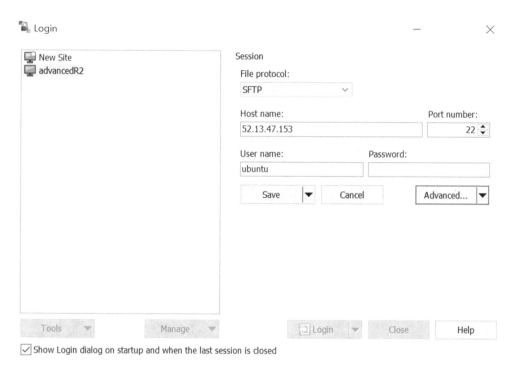

Figure 11-12. *WinSCP start screen with correct settings*

For the password, select Advanced, and in the new window that opens, in the SSH tab and under Authentication, navigate to the same **.ppk* file you used for PuTTY. We show the results in Figure 11-13. After clicking *OK* and then *Save*, you are ready to log in.

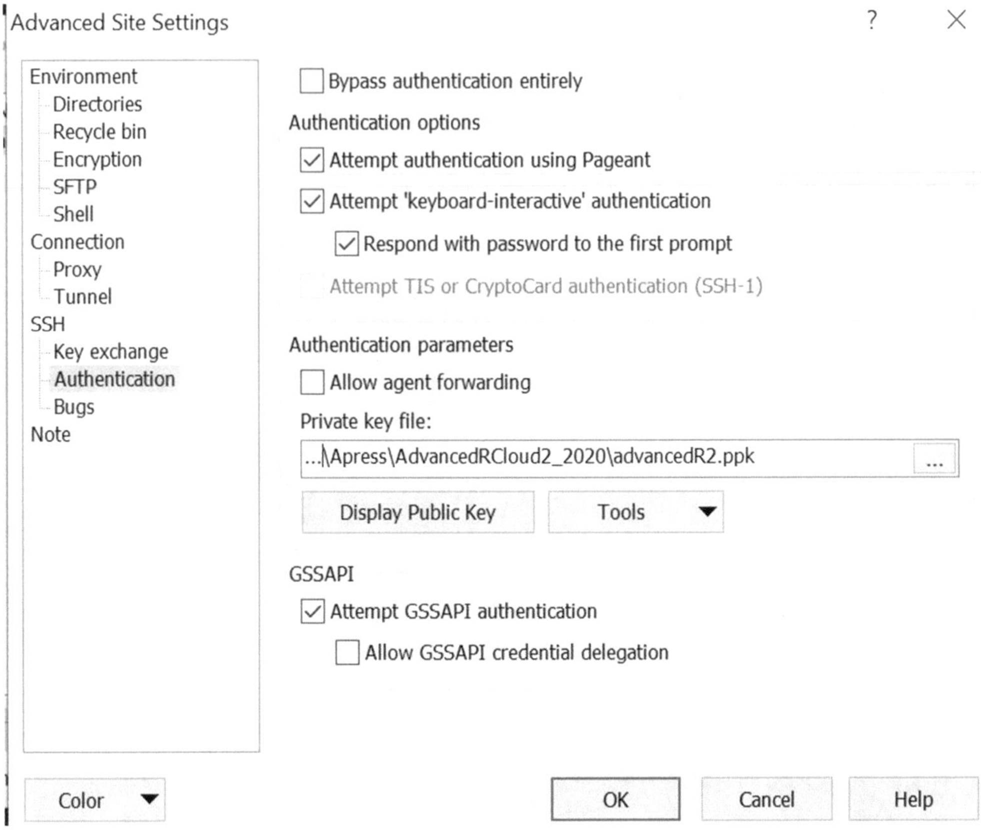

Figure 11-13. *WinSCP advanced key file screen*

When you saved, you had the option to create a desktop shortcut, which can be quite convenient if you regularly have files to transfer. At your first login, you see your server's key fingerprint (unless you loaded from PuTTY). Again, just as with PuTTY, be sure to check that against the fingerprint in Figure 11-8. We go ahead and upload that old Chapter 2 file called ch02_link.txt to our /home/ubuntu folder on our server. To do this, on the left side, navigate to where you are storing the files you downloaded from our online code packet. Then, drag and drop that text file to the right-hand side. Once it has uploaded, we go ahead and disconnect from the session and close out of the program. Remember the filename, though; we use that file in the next chapter.

11.6 Summary

Both PuTTY and WinSCP have versions that work without installation. Thus, even on a computer on which you do not have administrative rights, you can still access your instance and transfer files back and forth. This can be helpful in the corporate or academic environment, where technology services departments often rule local computer rights with an abundance of caution.

Speaking of caution, you should be cautious with your cloud server. While we built our EC2 instance with some thought to security, this is not entirely safe. Part of our goal is to get you using R on the cloud. Neither is our expertise network security. So while what we have shown here is likely somewhat safer than email, depending on your use case, you may need more. Often, research data may have either economic or privacy value. In that case, you undoubtedly want to consider some Internet security texts. Proceed at your risk; this is not a book about securing your cloud!

Rather than R functions only, Table 11-1 is a summary of new concepts or commands required to get your cloud instances/databases.

Table 11-1. *Listing of key concepts/commands described in this chapter and summary of what they do*

Concept/Command	What It Does
AWS	Amazon Web Services
PuTTY	A local terminal on Windows to securely connect to our cloud.
WinSCP	A local program on Windows to securely share files between our local machine and our cloud EC2 instance.
exit	An Ubuntu command-line command that exits the Unix terminal.

CHAPTER 12

Cloud Ubuntu for Windows Users

As you saw in the last chapter, using the cloud does not always allow for using Windows. This is not entirely true; Windows can certainly be utilized in the cloud. However, this tends to increase the cost in high-performance environments, making it impractical for some occasions. Additionally, it is not impossible to use the Ubuntu environment, and many of the same skills advanced R users have translate readily. All in all, using Ubuntu as your server instance operating system is not difficult with a little guidance to get started.

Our goals for this chapter include understanding the basic Ubuntu command line, updating and installing any operating system or packages required, installing both R and RStudio Server, and using R both on the command line and via the server. In particular, you will learn how to get an R process running on the server regardless of whether you are connected to that server. Thus, we can use PuTTY to check in as needed (along with AWS warning signals) to detect when a process has finished.

12.1 Getting Started

Provided you followed the last chapter through to the end, on your local machine's desktop, you should have links to both PuTTY and WinSCP. Additionally, both those programs have saved your session information.

To get started, open PuTTY, and use the saved information to access your cloud server. There will be a *Saved Sessions* box, and you can simply double-click the session you wish to start. If you are successful, you should see a screen much like Figure 12-1.

© Matt Wiley and Joshua F. Wiley 2020
M. Wiley and J. F. Wiley, *Advanced R 4 Data Programming and the Cloud*,
https://doi.org/10.1007/978-1-4842-5973-3_12

If you are not successful, there is one likely reason. Keeping in mind you did successfully connect in the prior chapter, most likely your IP address changed. Access AWS in a web browser, visit your *security group*, and check the inbound rules. Compare that with your current IP address. If you google "what is my ip," Google will tell you your address. If the IP address in the rules is different from your current IP address, make those changes, and save. Then try PuTTY again.

If you are still not successful, then the best troubleshooting advice we can give is to go step by step through the prior chapter. Somewhere along the way, something must have changed. There is no need to create a new EC2 instance most likely—simply review all the steps. The only reason why you might need to start fully from scratch would be if your key pair file was accidentally deleted.

12.2 Common Commands

Several of the commands we use in Ubuntu have counterparts in R. Others are more operating system specific rather than program specific. Many of the techniques we use to manage files in R can be done faster and more naturally via the command line. In this section, we explore the usual sorts of commands that are helpful to know and understand.

You are now in PuTTY, it looks about like Figure 12-1, and there is text blocking up our screen.

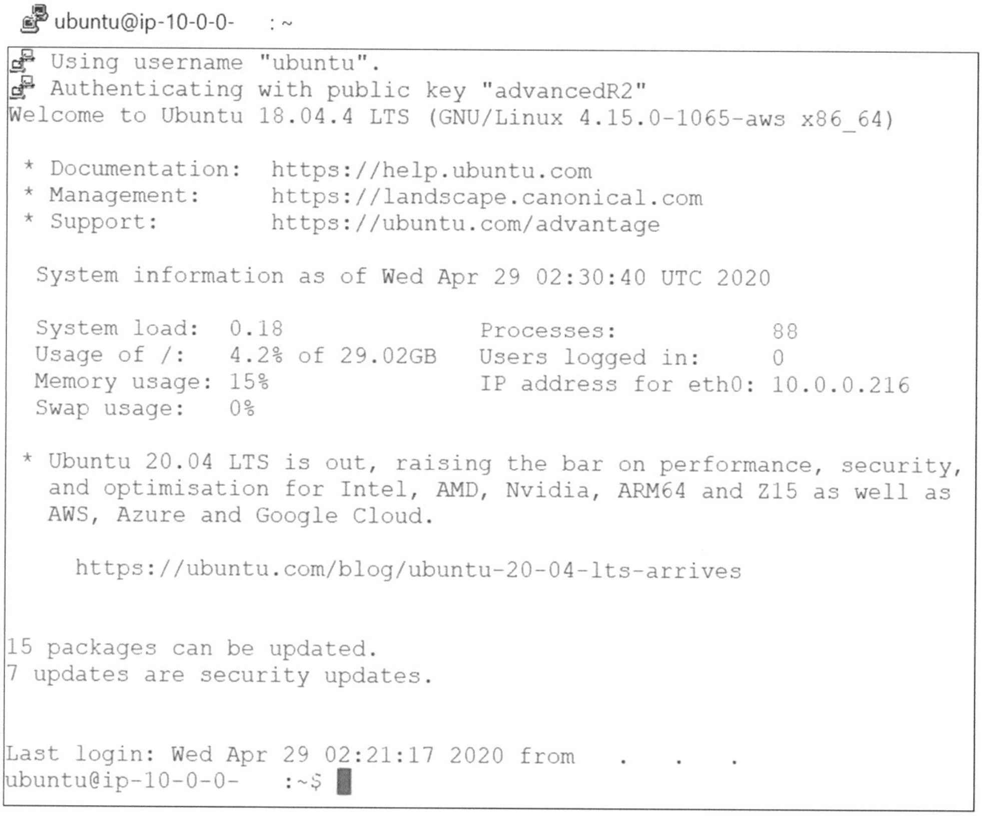

Figure 12-1. *The first screen shows seven security updates*

The first useful command is `clear`, a counterpart to `Ctrl+L` in R. This clears your viewing screen and allows you to easily see current commands and their results. We pair this command with `ls`, which is the list directory comments command. Right away the file we uploaded in the last chapter `ch02_link.txt` shows up.

Rather than survive only on screenshots, we will attempt to mimic the console output inline. In an Ubuntu console, much like in R, it is not enough to only type the code; one must press `ENTER` or `Return`. To minimize distractions, we will **not** show that `ENTER` command. Additionally, we will add $ to show what is typed at the console, although we will not include the entire username header:

$ **clear**

$ **ls**

ch02_link.txt

Now that we can see which files are in our directory, we can get a bit more information about those files. The `ls` command can be modified with several letters. The `-l` modification is the long format, and it gives information about the file rights (`-rw-rw-r--`), the user and the user group (both ubuntu, as we have a simple user set up currently), file size, as well as the date modified (notice this preserved our file's original modified date from before we uploaded it). Of particular note is the usage of file rights that relate to file security and safety. In Unix systems such as Ubuntu, all objects come with rights. The pattern is `-ooogggppp-`, where o refers to file owner, g relates to the group membership of the owner, and p is for public. Furthermore, each level of user can either *read*, *write*, or *execute* a particular object. This becomes quite important later, when we create some public-facing utilities that are accessible from the Web. It is important that we check the permissions, to ensure that the public can read files, yet not write or modify them. In our code that follows, both the user *ubuntu* and any future members of the group ubuntu are allowed both read and write access to our text file. The general public, if this directory were ever made public, would be able to only read our file:

```
$ ls -l
total 4
-rw-rw-r-- 1 ubuntu ubuntu 31 Mar 05:16 ch02_link.txt
```

We said the word *directory*, so we may as well see where we live in our system. Folders, or directories, also come with the same access permissions (they are objects), and we can create new directories and change directories. To present the working directory, we use pwd; to make a directory, we use `mkdir`; and to change the directory, we use cd. We show these commands and their results in the code that follows. Note that the `..` command returns us one level up in the directory structure, back to our original user directory:

```
$ pwd
/home/ubuntu

$ mkdir MyFolder
$ cd MyFolder
~/MyFolder$ pwd
/home/ubuntu/MyFolder

~/MyFolder$ cd ..
```

Having discussed making and navigating directories, we turn our attention to removing files. To remove a file, we use the command rm. We go ahead and do that to our text file that we uploaded; there was never any need for it other than as an example file. We show the removal in the following code, along with a full view of our user folder. Showing our list command with the full view allows us to see that our text file has been removed in the before and after views:

```
$ ls -hl
total 8.0K
-rw-rw-r-- 1 ubuntu ubuntu   31 Mar  5 02:30 ch02_text.txt
drwxrwxr-x 2 ubuntu ubuntu 4.0K Apr 11 21:21 MyFolder

rm ch02_link.txt
ls -hl
total 4.0K
drwxrwxr-x 2 ubuntu ubuntu 4.0K Apr 11 21:21 MyFolder
```

The values of these access permissions may be modified. The command chmod takes three number inputs after it in the pattern chmod ogp. We create a file and give three levels of access to the user, to the user's group, and to the public. Execute permission is given by 1, write is provided by 2, and read is given by 4. Additionally, numbers may be added to give more permission. See the example that follows: 7 = 1 + 2 + 4 gives the user full access, and 6 = 2 + 4 gives the group read and write access only. Finally, the public has only write access.

As we have no files in our database (we just deleted our text file), we can create a new file using the command touch. We call the new file permissionsFile:

```
$ touch permissionsFile
$ ls -hl
total 4.0K
drwxrwxr-x 2 ubuntu ubuntu 4.0K Apr 11 21:21 MyFolder
-rw-rw-r-- 1 ubuntu ubuntu    0 Apr 11 21:26 permissionsFile

$ chmod 762 permissionsFile
$ ls -hl
total 4.0K
drwxrwxr-x 2 ubuntu ubuntu 4.0K Apr 11 21:21 MyFolder
-rwxrw--w- 1 ubuntu ubuntu    0 Apr 11 21:26 permissionsFile
```

We end this section discussing two terminal commands, sudo and whoami. In the next section, we are going to start updating and maintaining our operating system. This requires root-level privileges. For security, our Ubuntu instance does not have a permanent root user, exactly. Instead, when we need elevated privileges, we give them to our ubuntu user by prefacing a command with sudo. This command stands for *super user do* and is always followed by another command. The system temporarily escalates privileges for the command that follows.

Notice in the first chunk, the system believes we are the default user. However, when using sudo, the system believes we are the root user:

```
$ whoami
ubuntu

$ sudo whoami
root
```

Now that we have the ability to run various commands as *root*, we turn to the next section, where we make sure that our cloud instance is up-to-date with the latest patches and packages.

12.3 Super User and Security

Just as in any operating system and software suite, often updates should occur. Some may add new or improved functionality, while others may be more security related. While comprehensive cloud and server security is far beyond the scope of this text, a few steps make sense to do directly. Looking at Figure 12-1, it is clear we have some work to do.

The update process involves three steps. We first run sudo apt-get update with the following abbreviated code output (we truncate with . . .):

```
$ sudo apt-get update
....
Fetched 3,451 kB in 3s (1,097 kB/s)
Reading package lists... Done
```

Next, we upgrade our current files so that we are ready to install new ones if we so desire. This takes some time, as we have quite a few to download! Again, we abbreviate some of the output. Of interest is to note that we did agree to continue when asked with a Y.

It is almost certain your specific number of upgrades and packages will be different from what is shown here. However, the commands you will use to update your cloud instance will stay the same:

```
$ sudo apt-get upgrade

Reading package lists... Done
Building dependency tree
Reading state information... Done
Calculating upgrade... Done
....
62 upgraded, 0 newly installed, 0 to remove and 4 not upgraded.
Need to get 23.5 MB of archives.

After this operation, 209 kB of additional disk space will be used.
Do you want to continue? [Y/n] Y
...
```

Now that our files are upgraded, we can install our new packages. We run the command sudo apt-get dist-upgrade at the command line, agreeing to the prompts when asked. While it is not necessarily required,we also reboot the system. This exits us from our session, and we have to re-access our cloud via PuTTY. We show both the abbreviated code to perform these commands and the Figure 12-2 screenshot showing no more updates possible:

```
    $sudo apt-get dist-upgrade
...

$sudo reboot
...
The system is going down for reboot NOW!
```

```
    ubuntu@ip-10-0-0-    : ~
   Using username "ubuntu".
   Authenticating with public key "advancedR2"
 Welcome to Ubuntu 18.04.4 LTS (GNU/Linux 4.15.0-1065-aws x86_64)

 * Documentation:  https://help.ubuntu.com
 * Management:     https://landscape.canonical.com
 * Support:        https://ubuntu.com/advantage

  System information as of Wed Apr 29 03:08:16 UTC 2020

  System load:  0.19              Processes:           90
  Usage of /:   9.5% of 29.02GB   Users logged in:      0
  Memory usage: 16%               IP address for eth0: 10.0.0.216
  Swap usage:   0%

 * Ubuntu 20.04 LTS is out, raising the bar on performance, security,
   and optimisation for Intel, AMD, Nvidia, ARM64 and Z15 as well as
   AWS, Azure and Google Cloud.

     https://ubuntu.com/blog/ubuntu-20-04-lts-arrives


 0 packages can be updated.
 0 updates are security updates.


 Last login: Wed Apr 29 02:37:48 2020 from
 ubuntu@ip-10-0-0-    :~$ █
```

Figure 12-2. *Ubuntu startup screen showing no needed updates or upgrades*

Depending on when you read this, AWS may or may not have updated the version of Ubuntu from version 18.04. It is not required to upgrade. In fact, it is usually more convenient to wait until Amazon takes care of such things in their initial instance setup. However, should you wish to run on the latest version, run the command sudo apt install update-manager-core followed by sudo do-release-upgrade. As of this writing, while Ubuntu 20.04 is released as a long-term stable version, it is not quite yet ready on AWS for an easy install. Most likely that will change by the time this book is in production.

We should note that also as of this book's release, R 4.0.0 is *not* fully supported for Ubuntu 20.04. While it most likely would work just fine, there would be some extra effort required to make everything work.

Should you choose to upgrade anyway, you will be warned that upgrading via remote access is not advised and that additional access SSH scripts are being run to ports that are not likely open to the public. In particular, our security settings would prevent access to this backup access port. Since you have just created this instance and can readily terminate from the AWS website, there is no need for such a backup access route. In general, we run such updates as available, agree to all the options when asked by typing y + ENTER, keep any local versions by pressing ENTER, and restart. Well, we say we run when available. In truth, it is worth having a couple days in a row to devote to any fixes that may be required. Most often that is not required. However, it is time worth having before you update your operating system.

If/when you do choose to upgrade the distribution, be sure to sudo update and upgrade as shown from the preceding commands one last time after getting your new operating system version.

Now that our instance or at least our packages on our instance are updated and we have a better sense of how to navigate in the Ubuntu file structure, we may turn our attention to installing R on our cloud server.

12.4 Installing and Using R

Our main goal so far in the last chapter and this one is to prepare a server to host R for us. We are very close to being there and have just a few more steps to perform. Our first step is to modify our package sources list so the Ubuntu operating system knows where to find R.

It will help to check your version of Ubuntu first. Notice our version as of this writing is 18.04 and code name is "bionic":

```
$ lsb_release -a

No LSB modules are available.
Distributor ID: Ubuntu
Description:    Ubuntu 18.04.4 LTS
Release:        18.04
Codename:       bionic
```

Allowing R to be found is achieved by changing to the /etc/apt/ directory and editing the sources.list file. Much like a PC has your text editor or Microsoft Word, Unix-style operating systems also have text editors. One of these is called "nano." Because sources.list lists the sources from which Ubuntu obtains packages and applications, it is a file that must be secured. Thus, sudo is called for, to escalate your privilege from the usual user:

```
$ cd /etc/apt/
/etc/apt$ ls
```

```
apt.conf.d      sources.list      trusted.gpg
preferences.d   sources.list.d    trusted.gpg.d
```

```
$ sudo nano sources.list
```

Now that the nano text editor is open, arrow down to the end of the file, press ENTER to add a new line, and type in **one** of the following lines of code. Then press Ctrl+O to save; you see your filename and hit Enter/Return. From there, press Ctrl+X to exit to the command line. Note *bionic* and *xenial* are the version 18 and 16 names for Ubuntu.

These next two options are key! You must choose **only one** option, depending on your Ubuntu version. Additionally, this will break as the new stable version of Ubuntu release after this book does (which of course it will). You will want to visit https:// cran.r-project.org/bin/linux/ubuntu/ and check for the latest *code name* for the latest version of both R and your Ubuntu and make the appropriate change in the following text to add to your source file. We give the last two versions to help you see the pattern should you need to make edits based on future software versions:

If you updated to Ubuntu 18.04 (or started there from AWS): Code name: bionic-cran35

```
##Add R File server
deb https://cloud.r-project.org/bin/linux/ubuntu bionic-cran40/
```

If you stayed with Ubuntu 16.04: Code name: xenial-cran35

```
##Add R File server
deb https://cloud.r-project.org/bin/linux/ubuntu xenial-cran40/
```

Remembering to use *Ctrl+O* and *Enter* to save, followed by *Ctrl+X* to exit, we move back into our home directory. Our next step is to add to our package utility this signature key for R:

```
E298A3A825C0D65DFD57CBB651716619E084DAB9}
```

This allows us to confirm downloads we make are likely safe to download and install. Please note this signature key may also change as versions of R change. A good place to check for updates or troubleshoot steps is `https://cran.r-project.org/bin/linux/ubuntu/`. If your installation has any errors, this and verifying your source file has the correct address are the top two places to double- and triple-check.

The signature key should be on *one line*. To fit in the book, we break after the space after the `--recv-keys`. We show both the commands you should type and the output you should expect:

```
$ cd /home/ubuntu
$ sudo apt-key adv --keyserver keyserver.ubuntu.com --recv-keys
E298A3A825C0D65DFD57CBB651716619E084DAB9

Executing: /tmp/apt-key-gpghome.rOCEL7Je1U/gpg.1.sh --keyserver keyserver.
ubuntu.com --recv-keys
gpg: key 51716619E084DAB9: public key "Michael Rutter <marutter@gmail.com>"
imported
gpg: Total number processed: 1
gpg:                   imported: 1
```

From here, installing R is (normally) simple. In fact, it should only take two commands. We update the packages list and then install base R. In the setup process, you need to agree to some options once or twice. If it is a successful installation, you should see a "wall" of text scroll past. If it stops moving, check to see if you are being asked to enter a y + ENTER to agree to something. Otherwise, let the installation complete:

```
$ sudo apt-get update
$ sudo apt-get install r-base
...
```

Now that we have installed R on our cloud, we run R and see what happens. We use the command R to start our program. We run one of our scripts from Chapter 1 to see what happens. The following code shows both the start screen of R and the familiar results of running our code:

```
$ R

R version 4.0.0 (2020-04-24) -- "Arbor Day"
Copyright (C) 2016 The R Foundation for Statistical Computing
```

```
Platform: x86_64-pc-linux-gnu (64-bit)
```

```
R is free software and comes with ABSOLUTELY NO WARRANTY.
You are welcome to redistribute it under certain conditions.
Type 'license()' or 'licence()' for distribution details.
```

```
R is a collaborative project with many contributors.
Type 'contributors()' for more information and
'citation()' on how to cite R or R packages in publications.
```

```
Type 'demo()' for some demos, 'help()' for on-line help, or
'help.start()' for an HTML browser interface to help.
Type 'q()' to quit R.
```

```
> x <- c("a", "b", "c")
> x[1]
[1] "a"
```

```
> is.vector(x)
[1] TRUE
```

```
> is.vector(x[1])
[1] TRUE
```

```
> is.character(x[1])
[1] TRUE
```

```
> q()
```

Using R

Inside R, all the commands we regularly use work as expected. We can install packages or run code. However, we would like to improve three aspects. The first is we are running R as a user, namely, the user ubuntu. It may well be that we wish other users of our server to be able to use R without needing to have sudo or root privileges. In that case, it may be helpful to install packages not just for our user, but for the entire machine. Second, running code from files that we have created on our local machines would be convenient. Third, our connection to our server lives at the mercy of the Internet. If there is a glitch, we would lose our work. Thus, it is also convenient to get to the point where we can set our R commands to run from the command line so that even if we disconnect, our server is still calculating away. This is especially helpful if you use this setup to throw a high-powered compute environment at a parallel processor-powered model.

To install packages for just a particular user, you use the `install.packages("")` function inside the R console as usual. However, to install a package for any user (including future users you may invite to your cloud), use the following command. What you need to do is pass a command from the command line to R. We also need to run this as root (to install for all users), and we need root privileges to do so, which `sudo` and `su -` provide. The `-c` option executes the code directly following it. Finally, we call R and set it to install the package:

```
$ sudo su - -c "R -e \"install.packages('pscore')\""
```

Suppose you wish to write some code on your local machine, yet then run that code on your cloud server. Back on your local machine, we go ahead and start RStudio:

```
options(width = 70, digits = 2)
```

Then, we might save a new file named `ch12_loopTest.R` that has only this bit of code in it:

```
x <- 1:1000

for(i in 1:length(x)){
  x[i] <- x[i]^3
}
head(x)

## [1]    1   8   27   64  125  216
```

Of course we know the expected results; we already saw essentially this from a prior chapter.

Next, we go to WinSCP and use that to connect to our ubuntu server (it is fine to keep PuTTY open). After connecting to your server, on the left or *local* side of WinSCP, navigate to your saved file `ch12_loopTest.R`, and drag it to the right or *cloud instance* side.

Now, return to PuTTY, and type `ls`:

```
$ ls

ch12_loopTest.R MyFolder permissionsFile
```

Our file is there! To run code already created, we use the command R CMD BATCH. To run and then see the results of our file, simply use this new command:

```
$ R CMD BATCH ch12_loopTest.R
$ ls
ch12_loopTest.R ch12_loopTest.Rout MyFolder permissionsFile
```

We can see the output in the file ch12_loopTest.Rout, and we can view it by using the nano editor we used earlier in this chapter. Remember that Ctrl+X exits from that editor. We show both the command to open the file and an abbreviated output to demonstrate the code ran successfully:

```
$ nano ch12_loopTest.Rout
....
> head(x)
[1] 1 8 27 64 125 216
>
> proc.time()
user system elapsed
0.156 0.038 0.201
```

We may now run R commands on our instance at will. This includes any parallel processing techniques you wish (keep in mind the free tier will not likely help you much with that). Provided our code is built not to require user input, it may be readily left to run. We do this with the screen command. Essentially, this opens another terminal from which we can detach. This allows that terminal to keep running even if our PuTTY terminal session ends (either because we wish it to end or because network vagrancies force our hand). We can see when our process is done running as well. The command we run after our screen command with modifiers is familiar; should you wish more details about screen, the command man screen gives plenty of information:

```
$ screen -A -d -m R CMD BATCH ch12_loopTest.R
$ screen -ls
There is a screen on:
        ip-10-0-0-244 (03/17/2020 04:43:40 AM) (Detached)
1 Socket in /var/run/screen/S-ubuntu.
$ screen -ls
No Sockets found in /var/run/screen/S-ubuntu.
```

As you can see from the second listing, no more processes are running. Thus, it is safe to access our output file to see the results. It is worth noting that most likely you will *not* see the first detached screen for shorter runs such as this one. It will simply be finished too quickly. For a process that took longer to run, rather than enter `screen -ls` to check up, we might just type `exit` and go about our day in the real world. The process will keep on running on the cloud. For processes that run on pricier instances, it is possible to set up AWS to email an alert when processor use drops below a certain threshold. We often use 10 percent or so. Once a couple alert emails have come that the process is staying below 10 percent, you know it is worth logging on to check.

We now have installed R on our server instance, can set up packages that are accessible to all users, and can run files without maintaining a steady network connection. If we design code on one machine that seems to take too long to run, we can upload those files to a more powerful machine and run them. However, we are still using the command line. In the next section, we install RStudio on our instance, so we can use the familiar look and feel of RStudio while still gaining the power of a cloud instance.

12.5 Installing and Using RStudio Server

While we have shown it is relatively straightforward to upload a file that was running too slowly on a local machine to our cloud instance, the format does leave something to be desired. RStudio [20] is a wonderful, intuitive environment in which to code R, and we can get that same graphic interface. What is more, this can be accessible from a browser rather than from the command line. This can be quite convenient, as just about any Internet-capable device with a browser can access your instance and use R. In a pinch, we used this once to run R via a smartphone.

The risk, of course, is some security concerns. RStudio Server is protected by only a password rather than a key file. Recall from prior chapters how many system administrative tasks, such as file moving and deletion, may be done by R. Access to RStudio Server may enable a dedicated attacker more access to your system than desirable. We, of course, continue to keep our security group on AWS set up to allow connections only from our local IP address, although this rather curtails the promise of use of R from any device. We remind you again that data security is worth a closer look than we give here. In the example we mentioned where we used a smartphone, we first had to access AWS, update the security group to allow the various connections from our phone's IP address, and then we got things running again. Without some other security measures, that would not have been the safest thing to do. On the other hand, safety vs. usability can often be a trade-off.

Since we have already installed R, we need only a little bit of work after accessing our command line again via PuTTY. To ensure proper package authentication, it helps to update, just as when we were installing system and security updates. After each of the following apt-get commands, much text prints to your console, and you may be prompted to agree to some disk space usage:

```
$ sudo apt-get update
...
$ sudo apt-get install gdebi-core
...
```

Go ahead and add RStudio's key to your machine. It is always possible RStudio might change their key. Visit `https://rstudio.com/code-signing/` to verify that `3F32EE77E331692F` is still correct. Additionally, we install `dpkg-sig` to help us verify our installation:

```
$ gpg --keyserver keys.gnupg.net --recv-keys 3F32EE77E331692F
...
$ sudo apt install dpkg-sig
```

Once you have the ability to install the server, you need to download it from the Web. Run the following command, and expect to download a file about 40 MB in size. We show the code and some of the output here; be sure to note the saved name of the file as well as a full download being complete. Additionally, be sure to install the correct version for your version of Ubuntu. We again show the code names to help you see the pattern as it is likely by the time you read this that there will be a newer version of at least one of Ubuntu or RStudio Server (likely both). Visit `https://rstudio.com/products/rstudio/download-server/debian-ubuntu/` for the latest details.

If you updated to Ubuntu 18.04 (or started there from AWS): Code name: `bionic`

```
$ wget https://download2.rstudio.org/server/bionic/amd64/rstudio-server-1.2.5042-amd64.deb
...

$ dpkg-sig --verify rstudio-server-1.2.5042-amd64.deb
```

```
Processing rstudio-server-1.2.5042-amd64.deb...
GOODSIG _gpgbuilder FE8564CFF1AB93F1728645193F32EE77E331692F 1585759224
```

If you had Ubuntu 16.04: Code name: xenial / (trusty is an older version)

```
$ wget https://download2.rstudio.org/server/trusty/amd64/rstudio-server-
1.2.5033-amd64.deb
...

$ dpkg-sig --verify rstudio-server-1.2.5033-amd64.deb

Processing rstudio-server-1.2.5033-amd64.deb...
GOODSIG _gpgbuilder FE8564CFF1AB93F1728645193F32EE77E331692F 1575443526
```

In either case, you got the GOODSIG, and thus you have a safer download.

Now we install the server. This should require one yes agreement during the installation process:

```
$ sudo gdebi rstudio-server-1.2.5042-amd64.deb

Reading package lists... Done

...

Do you want to install the software package? [y/N]: \textbf{y}

...

    Active: active (running) since Sun 2020-04-26 18:07:02 UTC; 1s ago
   Process: 26414 ExecStart=/usr/lib/rstudio-server/bin/rserver
(code=exited, status=0/SUCCESS)
  Main PID: 26420 (rserver)
     Tasks: 3 (limit: 1152)
    CGroup: /system.slice/rstudio-server.service
            |--26420 /usr/lib/rstudio-server/bin/rserver

Apr 26 18:07:02 ip-10-0-0-    systemd[1]: Starting RStudio Server...
Apr 26 18:07:02 ip-10-0-0-    systemd[1]: Started RStudio Server.

...
```

Now that RStudio Server is running, we want confirmation that it is properly working. Remember in the last chapter we set our security group to allow access to RStudio Server on port 8787. Open a web browser on your local computer to your server's web address at the needed port, namely, 54.201.117.64:8787 (or, for you, YOUR_ INSTANCE_IP_HERE:8787).

As you can see in Figure 12-3, it does indeed visually appear that the server is working. If it does not seem to work for you, please remember to check in the AWS website that your *current* IP address is the same as the IP address for the inbound rules of your security group. Details are outlined in Chapter 11.

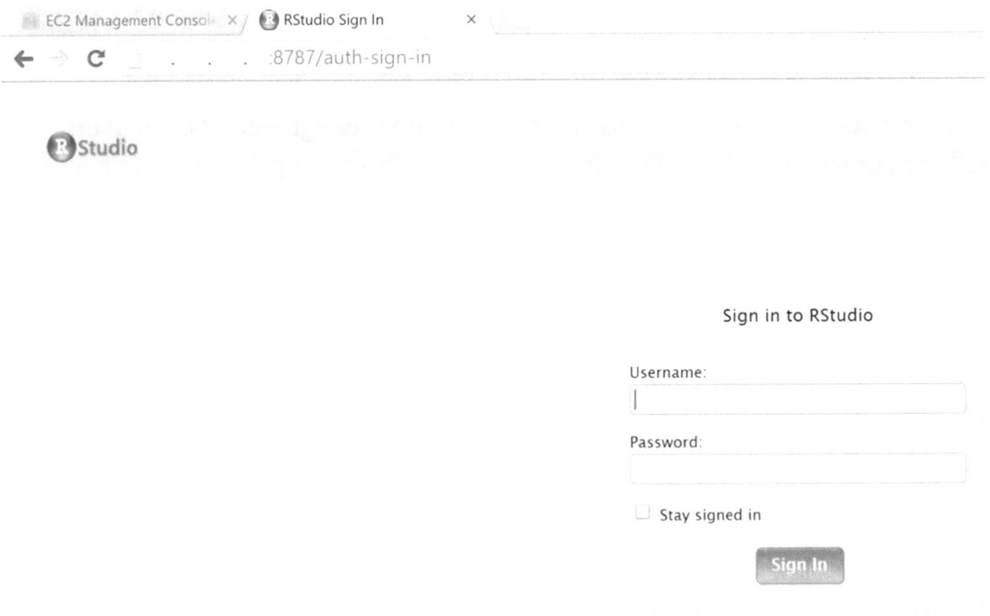

Figure 12-3. RStudio Server

We need a username and password to gain further access. The user is also a user of your Ubuntu server and has a password. It is wise to have a password log on to the server disabled by default, which our instance does in fact already have. The command we use to create a new user is adduser, and this runs the script shown for us. We need to type these commands back in our command line. It is not necessary to fill in information for your new user beyond the username and password. You can simply press ENTER to move to the next line. Also, note the password field has suppressed keystrokes so that nothing populates there. Choose any secure password you like, although we recommend you store it in some way. We used the *in*secure password: secure. Do not do this! Make your password secure (we felt safe enough as we are using a toy instance for this book that will soon be deleted).

We go back to our PuTTY terminal and type our adduser command:

```
$ sudo adduser advancedr

Adding user 'advancedr' ...
Adding new group 'advancedr' (1001) ...
Adding new user 'advancedr' (1001) with group 'advancedr' ...
Creating home directory '/home/advancedr' ...
Copying files from '/etc/skel' ...
Enter new UNIX password:
Retype new UNIX password:
passwd: password updated successfully
Changing the user information for advancedr

Enter the new value, or press ENTER for the default
        Full Name []: Advanced R
        Room Number []: RStudio Server
        Work Phone []:
        Home Phone []:
        Other []:
Is the information correct? [Y/n] Y
```

Going back to our browser, we use our new username and password to access RStudio Server. Enter the username and password just created, as in Figure 12-4, and you should be into RStudio on the Web.

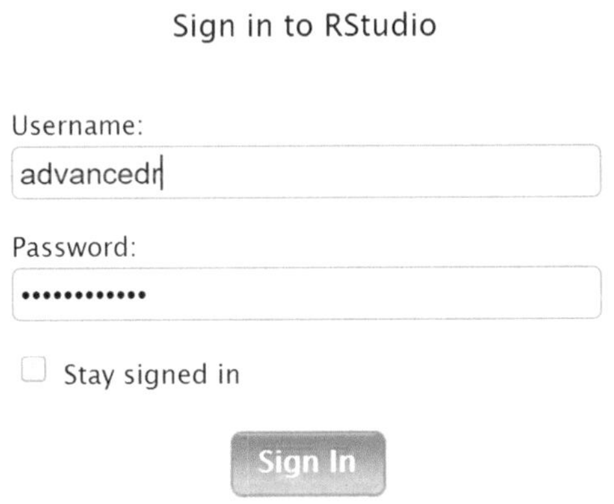

Sign in to RStudio

Username:

advancedr

Password:

••••••••••••

☐ Stay signed in

Sign In

Figure 12-4. *Sign into RStudio*

This should feel just like the RStudio we have been using all along, as shown in Figure 12-5.

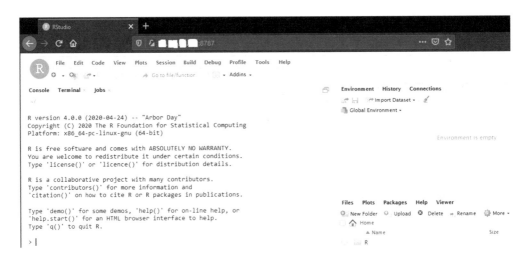

Figure 12-5. *RStudio on your cloud*

RStudio now exists on your cloud instance. While this is helpful for a more intuitive way to work with R, the ability to more natively use and see graphical output is something we, the authors, very much appreciate. Such output may be readily saved to *.pdf files and downloaded via WinSCP from the /home/advancedr folder. More generally, /home/YOUR_USERNAME accesses a particular RStudio user folder.

This completes our setup of RStudio Server. More could be done, but this is enough for starters. You now have access to a potentially powerful system and indeed are free to have several duplicates of such systems if you wish. On AWS, it is possible to set up warnings that alert you via email when your instance(s) drops below a certain CPU utilization rate. Thus, an R process may be configured to run, and instance space need not be paid for much beyond actual use time. Our final suggestion is to spend some time with the RStudio Server documentation if you intend to use it more than occasionally. There are some useful customizations, convenient features, and good-to-know caveats. We include the link in the reference section of this chapter.

12.6 Installing Java

Next, we install Java. It is quite easy; just run the following three lines of code, one at a time, in the command line. Agree to things as needed with y + ENTER:

```
$ sudo apt-get update
...

$ sudo apt-get install default-jre
...

$ sudo apt-get install default-jdk
...
```

12.7 Installing Shiny on Your Cloud

Bringing statistics to information consumers is the main use for Shiny. Because we are in installation mode, we'll go ahead and install the needed resources on our cloud instance to host a Shiny application.

To install Shiny, we need to run the following code. An agreement is required to confirm an installation, and we suppress the system's output to the command line. Run each line by itself.

This line installs the shiny package [6] into R. We also install httpuv [44] and shinydashboard [5]. You notice that between the inner parentheses, there are commands that make lots of sense to R users. However, in this case, we are running R from the command line, specifically, so we can execute the process with administrative privileges and thus have access to these libraries from any Ubuntu user account rather than just the main user account. When a Shiny server hosts applications, it hosts them by using the username shiny. Hence, the following lines of code install these packages for all users:

```
$ sudo su - -c "R -e \"install.packages('httpuv')\""
...
$ sudo su - -c "R -e \"install.packages('shiny')\""
...
$ sudo su - -c "R -e \"install.packages('shinydashboard')\""
...
```

A note of warning: This latest edition of R seems to not always install packages quite as it ought on Ubuntu micro-instances. You may, upon attempting to install one of the preceding packages, find that the system freezes. If that is the case, it may become necessary to forcibly reboot your instance in the AWS EC2 Dashboard.

Near where you would have launched your instance, there is an *Actions* drop-down. If you first select your instance, then *Actions*, and then *Instance State* followed by *Reboot*, your instance should reboot. If that does not work, you may need to *Stop* and then *Start* your instance. In that case, please note your instance's IP address will change, so PuTTY would need to be updated with that new IP address.

Upon a successful reboot or restart, and using PuTTY to get back into your command line, you may find that a *lock* is on one or more of your packages. Change to your R directory, and use ls to see if there are any folders listed in the format 00LOCK-rPackageHere. We show in the following how that may look:

```
$ cd /usr/local/lib/R/site-library/
$ ls
```

```
00LOCK-httpuv
```

If you do see such a lock on a file, you can remove the file by using the following command:

```
$ sudo rm -r 00LOCK-httpuv/
```

Be cautious with this command. You **must** be sure to type it precisely; the recursive remove command can easily remove far too many files. Once you have removed all folders that have the prefix OOLOCK-, you are ready to return to your home directory and try again. One technique that may help your retry work better is to research package dependencies on CRAN. For example, httpuv is a package on which shiny depends. Because we found that package especially fragile on a micro Ubuntu instance, we went ahead and had you install it separately:

cd /**home**/ubuntu/

From here, now that R itself can use Shiny, we install the server so our cloud instance can serve Shiny apps to the Internet. We have to download a file with the wget command and then install the downloaded file. We do that with the following lines of code, and there are some installation disclaimers. Additionally, we show an example of dpkg-sig showing no signature. Even good efforts at being secure fail on occasion:

```
$ wget https://download3.rstudio.org/ubuntu-14.04/x86_64/shiny-server-
1.5.13.944-amd64.deb
. . .

$ dpkg-sig --verify shiny-server-1.5.13.944-amd64.deb

$ sudo gdebi shiny-server-1.5.13.944-amd64.deb
...
Do you want to install the software package? [y/N]:\textbf{y}
...
. shiny-server.service - ShinyServer
    Loaded: loaded (/etc/systemd/system/shiny-server.service; enabled;
    vendor preset: enabled)
    Active: active (running) since Sun 2020-04-12 22:39:18 UTC; 67ms ago
Main PID: 18780 (shiny-server)
    Tasks: 1 (limit: 1151)
   CGroup: /system.slice/shiny-server.service
       |--18780 /opt/shiny-server/ext/node/bin/shiny-server /opt/shiny-server/
              lib/main.js

Apr 12 22:39:18 ip-10-0-0-    systemd[1]: Started ShinyServer.
```

Shiny Server should be running now too.

12.8 PostgreSQL

Previously, we learned how to install and run a PostgreSQL database server on our local machine. While that could be useful enough to help store large amounts of data from R in an organized fashion, it is not leveraging the full, community applications of a database server. In particular, our example mentioned creating several graduate students.

Run the following commands (which look fairly familiar by now), and agree when asked if you wish to continue. This will install PostgreSQL on your EC2 instance:

```
$ sudo apt update
$ sudo apt install postgresql postgresql-contrib
```

By default, PostgreSQL will only run on the local machine. While it starts to sound tired (and we promise we will finish such comments with this chapter), it should be repeated that having a database open to the world via the Internet is inviting trouble. It is definitely worth taking time to explore ways of securing your cloud. We have taken and will take some early steps that are likely as safe (or just a touch safer) than emailing information. Because each use case is different (and because this book is meant to focus on R and letting you practice that software), if your data requires more sensitivity, your first order of business ought to be to explore databases and cloud networking security.

To allow PostgreSQL to connect to the Internet, we need to add a line in the configuration file. Using nano, access the configuration file, and arrow down to the *CONNECTIONS AND AUTHENTICATION* section. Then, add listen addresses = '*'. This is a wildcard and will listen to *all* web addresses. However, this database server lives on your EC2 instance and is protected by your security group inbound filter setting which (currently) only allows your IP address (or someone faking/spoofing that exact address) to access information. Thus, the wildcard, wide-open choice here is not quite as dangerous as it may seem. We also added a comment to the side. Again, use Ctrl+O to save along with ENTER. Finally, use Ctrl+X to exit nano:

```
$ sudo nano /etc/postgresql/10/main/postgresql.conf

...
#------------------------------------------------------------------------

# CONNECTIONS AND AUTHENTICATION

#------------------------------------------------------------------------
```

```
# - Connection Settings -

#listen_addresses = 'localhost'       # what IP address(es) to listen on;
listen_addresses = '*'                # added for web access; ip filter via aws!
```

Next, we create a new user with full privileges named adminaws. Remember, PostgreSQL by default creates a super user named postgres. However, by keeping that user with **no** password, we force that user to only connect from the Ubuntu command line via sudo. This authenticates via your *key* and is thus much tougher to brute force.

Security is always about least privilege. What is the lowest amount of access that allows a particular user to achieve their goals? The balance is with convenience. Since our goal is to teach you how to use R to connect to SQL databases, we need most things to work easily on the first try, rather than falling down while nuances are sorted. Thus, this new user will have many privileges, including a password (P), the ability to create new roles (r), and super user rights (s). We kept the advancedr password as well:

```
$ sudo su - postgres -c "createuser -Prs adminaws"

Enter password for new role:
Enter it again:
```

Next, we create a database on our server named advancedrdb:

```
$ sudo su - postgres -c "createdb advancedrdb"
```

With a new user and a new database, we must give that new user access to that database. To do that, we first must use sudo to access the SQL database directly from the Ubuntu command line (recall our default user has no password). Once inside PostgreSQL's console, we use an old, familiar bit of SQL code to GRANT access. Then we use a new bit of code, \q, to quit from PostgreSQL:

```
$ sudo -u postgres psql

psql (10.12 (Ubuntu 10.12-0ubuntu0.18.04.1))
Type "help" for help.

postgres=# grant all privileges on database advancedrdb to adminaws;
postgres=# \q
```

We promise you are very close to seeing how to access a cloud SQL database! Hang in there.

The last thing that is required is to give access to the database in another configuration file. Again using nano, we arrow down through the file. We intend to add one line now to the very end of that file.

The line we are adding is host all all 0.0.0.0/0 password.

Notice that, with more lines, we could be more specific in which databases (e.g., advacedrdb only perhaps) or which users (e.g., gradstudentr only perhaps) or that the IP address could not be open to the entire world. For now, and again keeping in mind IP addresses are restricted (and we only have a handful of users), all may well be safe enough:

```
$ sudo nano /etc/postgresql/10/main/pg_hba.conf

...
# TYPE   DATABASE         USER              ADDRESS                    METHOD
# all users can access all databases from all locations via password.
host     all              all               0.0.0.0/0                  password
```

With all these configuration file changes, we had best restart our PostgreSQL server. In general, servers need to be restarted for changes to happen:

```
$ sudo service postgresql restart
```

While we do not repeat all of Chapter 10 here, we show just enough to convince you that you can now practice connecting to cloud-based SQL servers via R:

```
library(RPostgreSQL)
library(DBI)
drv <- dbDriver("PostgreSQL")

conSuper <- dbConnect( drv,
                dbname = "advancedrdb",
                host = "52.13.47.153",
                port = 5432,
                user = "adminaws",
                password = "advancedr"
                )
```

```
dbGetInfo(conSuper)

## $host
## [1] "52.13.47.153"
##
## $port
## [1] "5432"
##
## $user
## [1] "adminaws"
##
## $dbname
## [1] "advancedrdb"
##
## $serverVersion
## [1] "10.0.12"
##
## $protocolVersion
## [1] 3
##
## $backendPId
## [1] 1643
##
## $rsId
## list()

dbDisconnect(conSuper)

## [1] TRUE

rm(conSuper)
```

We close this section by noting that everything in that prior chapter should work. Thus, if this is a skill you seek, you are well on your way.

12.9 Summary

You now have access to a cloud instance that has RStudio Server (which can be quite convenient). In particular, you have the ability to upload files to and from a local machine in a way that does not require super user local privileges. If you opened up permission access by unrestricting more IP addresses, this is a way to use R from a smartphone. Also, we installed several packages and applications on our Ubuntu server. In the next chapters, we use those features to make some engaging analytics happen. More important, because they are served on the cloud, these data and results are accessible to anyone. Thus, sharing results has never been easier.

Table 12-1 references Ubuntu commands that are helpful.

Table 12-1. *Listing of key functions described in this chapter and summary of what they do*

command	What It Does
clear	Clears the console, like Ctrl+L in R.
ls	Lists files in the current directory that can be modified by -hl (see man ls for details).
pwd	Present working directory.
mkdir	MaKe DIRectory.
cd	Change directory.
touch fileName	Creates a file with the given name.
chmod	Changes/modifies file permissions for read/write/execute access.
whoami	Identifies current user.
sudo	Super user do.
sudo apt-get update	Updates list of packages.
sudo apt-get upgrade	Upgrades the files on hand.
sudo ap-get dist-upgrade	Installs the upgrades to the current file distribution.

(continued)

Table 12-1. (*continued*)

command	What It Does
sudo reboot	Reboots your cloud instance. It will take a bit before you can reconnect.
lsb_release -a	Shows current version of Ubuntu.
nano	A text editor.
R CMD BATCH fileName.R	Runs a .R file from the command line, without opening.
screen -A -d -m R CMD BATCH fileName.R	Runs a .R file from a new screen line, without opening in R, and you could exit and your file would still be running away in the cloud until it was complete.
screen -ls	Lists the current screens; if a process is currently running, it will show one or more sockets.
wget	web get—downloads a file to the current directory.

CHAPTER 13

Every Cloud Has a Shiny Lining

When serving data to the public, easy access and interactive information make a real difference in comprehension. In this chapter, our goal is to provide access and some interesting uses for a more recent application, shiny [6].

Shiny is a web application framework for R. What it does is allow information consumers to interact with live data and analytics in R. This framework does not require knowledge of any web-based languages. All the same, surprisingly complex and interactive data models can be released to your clients, decision makers, or consumers.

Do you have data your customers should explore? Are there performance metrics that, transformed from static to interactive, might support your story to a board or a boss? Do your users have data they need to upload and understand better? Would you like to easily update next quarter's data into a common dashboard? All these questions can be answered with shiny server. If we can get Shiny onto our cloud, then anyone can benefit from the power of R.

Now, for simplicity's sake, in this chapter we are going to live almost entirely on Windows inside RStudio. At the very end, we will host some stand-alone shiny applications on our server. In the next chapter, we put all the building blocks we had in this chapter into one cloud-hosted dashboard.

You will need to ensure your local computer has shiny [6] using install. packages("shiny") if needed:

```
options(width = 70, digits = 2)
library(knitr)
library(shiny)
```

M. Wiley and J. F. Wiley, *Advanced R 4 Data Programming and the Cloud*,
https://doi.org/10.1007/978-1-4842-5973-3_13

13.1 The Basics of Shiny

Think of shiny as a web interface for R code. If you can do it in R, Shiny can push those results to a nice page viewable via Internet browsers. Even more awesomely, if a data consumer wants to interact with information, their *input* can be incorporated into R code.

In RStudio, select *File, New File*, and *Shiny Web App*. Name your file PieChart_Time, and under Application choose the single file option (app.R). Delete all code and type in the following code (or download from this book's code repository/Apress site). As you type it in, you want to notice three things. First, there is a user interface portion of the code. Second, there is a server side to the code. Finally, the application must be run. Another aspect to notice (and this is why we recommend typing it all in manually) is where pieces of the user side and the server side mesh. In fact, in our code, there are exactly four places this happens. From the user side, variables of a number between 0 and 100, a text string, and a check box are *input*. The server receives those values and creates plots. It puts those plots into the *output*. The user side gets that output in mainPanel():

```
library(shiny)

# user interface drawing a pie chart
ui <- shinyUI(fluidPage(

   # Page title
   titlePanel("User Controlled Chart"),

   # Sidebar with numeric, text, and check box inputs controlled by user
   sidebarLayout(
      sidebarPanel(
         numericInput("pie",
                     "Percent of Pie Chart",
                     min = 0,
                     max = 100,
                     value = 50),

         textInput("pietext", "Text Input", value = "Default Title",
                     placeholder = "Enter Your Title Here"),
```

```
        checkboxInput("pieChoice",
                      " I want a Pie Chart instead.", value = FALSE)
    ),

    # Show the plot(s)
    mainPanel(
      plotOutput("piePlot")
    )
  )
))

# server side R code creating Pie Chart or barplot hidden from user
server <- shinyServer(function(input, output) {

  output$piePlot <- renderPlot({
    # generate Pie chart ratios based on input$pie from user
    y <- c(input$pie, 100-input$pie)

    # draw the pie chart or barplot with the specified ratio and label

   if(input$pieChoice == FALSE){
     barplot(y, ylim = c(0,100),
             names.arg = c(input$pietext, paste0("Complement of ",
             input$pietext)))
   }else{
     pie(y, labels = c(input$pietext, paste0("Complement of ",
     input$pietext)))}
  })
})

# Run the application
shinyApp(ui = ui, server = server)
```

Now, go into your app.R file you carefully typed and run that code in one block. There should even be a *Run App* play button near the top right of the code area. Either way, something along the lines of Figure 13-1 should appear.

Go ahead and enter various numbers and text values. Play with `shiny` and see what it can do. While you do so, recognize that what Shiny has done is provide an interface between you and R. Now, this particular run is hosted by RStudio on your local machine. Notice in the console it indicates listening, and you may need to select a *Stop* button before you can enter any new code.

So what happened? Suppose you changed the *Percent of Pie Chart* number to 75. Well, `numericInput()` is a reactive value. When you change it, behind the scenes, a signal is sent alerting the server that `input$pie` changed. The server then notifies any reactive functions using `input$pie` that something has changed. In our case, `renderPlot()` is a reactive function. It now knows `input$pie` is different, so it queries that value, gets the new value, and runs all its code again. Since that is an *output* value, this whole cycle repeats in reverse with `output$piePlot`, with the result that `plotOutput()` on the user side refreshes.

There are some important things to notice about all this. The process we just described runs all the code again inside `renderPlot()`)—all of it. Now, all else equal, there is no need to look up `input$pietext` and `input$pieChoice`. Our reactive function knows its values, for those are up-to-date, and while they are used again, they are not called again. Still, if we had something a little more intensive than a subtraction and some plots inside `renderPlot()`, we might find ourselves with a relatively steep performance cost to making any changes. While this set of code is quite compact, nevertheless you can see three places where you might put R code. One is where `library(shiny)` is. This is on the outside of our `shinyServer()`. Code that is here, such as our library call, is run **once** each session. A single session can have more than one user interacting with it (although that gets a bit esoteric at the moment). Still, each new user executes all the code inside the `shinyServer()` at least once. Finally, as we just saw, the code inside reactive functions such as `renderPlot()` are run once each update from any connected reactive input values. What this means is that for efficiency, consider placing as much code as possible outside the reactive functions. Moreover, if you anticipate heavy user traffic, place as much outside the server as possible too.

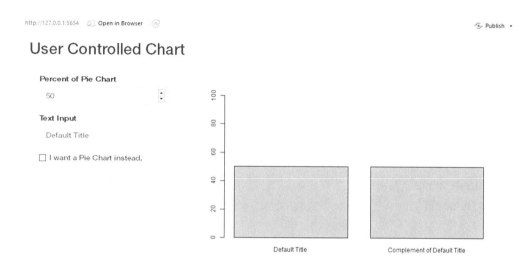

Figure 13-1. *A locally hosted Shiny app to see just what we are gaining*

For now, we introduce some new functions. Remember, there is the user input side, there is the server side, and there is the entire shiny app. Deep down inside, shiny is just those three pieces. In fact, here is a rather bare-bones application:

```
library(shiny)
# Define user interface
ui <- shinyUI(fluidPage())

# Define server
server <- shinyServer(function(input, output) {})

# Run the application
shinyApp(ui = ui, server = server)
```

You can run this code and get a nice blank page. While that is perhaps less inspiring than the previous code, this helps you get coding in this interface quickly. In the user interface section, what we see is the fluidPage() function call, inside of which we place all our user-facing text. This area of the code is the most unfamiliar part of shiny, because this is the code that creates the HTML seen by users. In the reference section, we place links to all the Shiny UI functions. However, for this chapter, we keep to the same basic layout inside a page that is fairly self-adjusting, or fluid, depending on the browser viewing area. Notice this is an R function, and functions in R take arguments that are separated by commas. Every aspect of our growing application is contained inside this wrapper function, which takes care of all the conversion to web-ready markup. For our first app from Figure 13-1, we used a sidebar layout. This function, in turn, wants two arguments: the sidebar panel where we control our application and then the main panel where we draw it. Editing our user interface gives us the following code:

```
# Define user interface
ui <- shinyUI(fluidPage(
    sidebarLayout(
        sidebarPanel(),
        mainPanel()
    )
))
```

This code is a wonderful example of the power of the mathematical notion of function composition. Our goal is to eventually have something the user inputs to our R code, such as the text or check box of our example in Figure 13-1.

Figure 13-2. *A mostly blank Shiny app, with a sidebarPanel()*

Figure 13-3. *The user interface is almost done, and the input portion in the sidebar is indeed done*

The user inputs those values into as-yet-to-be-named functions, which give visual output. That output in turn is input into the `sidebarPanel()` function, which provides web output, which is input to `fluidPage()`, which outputs the final web user interface. We can see the results, blank though they be, in Figure 13-2.

From this point, we can quickly build our user's inputs with three input functions. Each of these input functions follows a specific format, namely, *blank*Input(input identification, text label for user readability, settings relating to the specific type of input for *blank*). Formally, our three input functions have these layouts, and please notice the similarities!

```
numericInput(inputId, label, value, min = NA, max = NA, step = NA, width = NULL)
textInput(inputId, label, value = "", width = NULL, placeholder = NULL)
checkboxInput(inputId, label, value = FALSE, width = NULL)
```

Proceeding to fill in some specifics for our code, we can set the input identities for each to be useful variables for us to use later in our server code. Since we do plan to build a pie chart (and please, gentle readers, do not hate us for that choice), it makes sense to limit our numeric input to between 0 and 100. Because our server code runs once on start, it also makes sense to provide a default value for the numeric input and, indeed, the text input. Building further on our burgeoning user interface code, we now have this:

```
# user interface drawing a pie chart
ui <- shinyUI(fluidPage(
  # Sidebar with numeric, text, and check box inputs controlled by user
  sidebarLayout(
    sidebarPanel(
      numericInput("pie", "Percent of Pie Chart",
                   min = 0, max = 100, value = 50),
      textInput("pietext", "Text Input", value = "Default Title",
                placeholder = "Enter Your Title Here"),

      checkboxInput("pieChoice", " I want a Pie Chart instead.", value = FALSE)
              ),

    mainPanel()
  )
))
```

Again, after making those modifications, it makes sense to run the whole app again, and we see the new result in Figure 13-3. Notice that the mainPanel() is currently empty in our code, and, indeed, in Figure 13-3 we do not see anything on the right-hand side.

Now that we have our user input captured, we are ready to use it on the server side. Remember, shiny is an R wrapper that takes input and, when that input changes, alerts a reactive function on the server side that something is different. Behind the scenes, shiny is taking care of knowing when the input changes and alerting the correct reactive functions on the server; we need not worry about such things. Our server defines a new function that takes both input and output as formals and uses those in its body. If needed, take a quick look back at Chapter 4 to refresh your memory on just what a function is. Since all we want to do is draw either a bar plot or a pie chart, the majority of our code does precisely that in very common R code. In fact, the only part that is not fairly normal R code is our reactive function renderPlot(). Since we wish to give graphical outputs, we use that function.

Reactive functions are always waiting for shiny to alert them to any changes in the inputs from the user interface side. When an input changes, the reactive functions are notified if they have at least one of the changed values in their function body. If they do, then the reactive functions run their code again, this time with the new value(s). We see this in the following code. The only other part of this is that the output of renderPlot() is stored in an output variable, and we choose that name to be piePlot (even though it might not output a pie chart, depending on the check box choice—which is wholly our fault for selecting a poor name):

```
# server side R code creating Pie Chart or Bar plot
server <- shinyServer(function(input, output) {

    output$piePlot <- renderPlot({
        # generate Pie chart ratios based on input$pie from user
        y <- c(input$pie, 100-input$pie)

        # draw the pie chart or barplot with the specified ratio and label

      if(input$pieChoice == FALSE){
        barplot(y, ylim = c(0,100),
                names.arg = c(input$pietext, paste0("Complement of ",
                input$pietext)))
      }else{
        pie(y, labels = c(input$pietext, paste0("Complement of ",
        input$pietext)))}

    })
})
```

That is essentially it. When our server runs, it alerts the user interface function plotOutput("piePlot") to the new image. This function lives in the mainPanel() region. Under this simple model of shiny, the sidebar is where user input is received, the server side is where it is processed behind the scenes by R, and the main panel is where the processed output is viewed by the user. Just this once, we show the full HTML code that Shiny creates for us to serve in Figure 13-1.

```
<!DOCTYPE html>
<html>
  <head>
    <meta http-equiv="Content-Type" content="text/html; charset=utf-8"/>
    <script type="application/shiny-singletons"></script>
    <script type="application/html-dependencies">
json2[2014.02.04];jquery[3.4.1];shiny[1.4.0.2];bootstrap[3.4.1]
    </script>
    <script src="shared/json2-min.js"></script>
    <script src="shared/jquery.min.js"></script>
    <link href="shared/shiny.css" rel="stylesheet" />
    <script src="shared/shiny.min.js"></script>
    <meta name="viewport" content="width=device-width, initial-scale=1" />
    <link href="shared/bootstrap/css/bootstrap.min.css" rel="stylesheet" />
    <script src="shared/bootstrap/js/bootstrap.min.js"></script>
    <script src="shared/bootstrap/shim/html5shiv.min.js"></script>
    <script src="shared/bootstrap/shim/respond.min.js"></script>
    <title>User Controlled Chart</title>
  </head>
  <body>
    <div class="container-fluid">
      <h2>
        User Controlled Chart
      </h2>
      <div class="row">
        <div class="col-sm-4">
          <form class="well">
            <div class="form-group shiny-input-container">
              <label class="control-label" for="pie">
```

```
                Percent of Pie Chart</label>
                <input id="pie" type="number" class="form-control" value="50"
                min="0" max="100"/>
            </div>
            <div class="form-group shiny-input-container">
                <label class="control-label" for="pietext">
                Text Input</label>
                <input id="pietext" type="text" class="form-control"
                value="Default Title" placeholder="Enter Your Title Here"/>
            </div>
            <div class="form-group shiny-input-container">
                <div class="checkbox">
                    <label><input id="pieChoice" type="checkbox"/>
                    <span>I want a Pie Chart instead.</span></label>
                </div>
            </div>
          </form>
      </div>
      <div class="col-sm-8">
        <div id="piePlot" class="shiny-plot-output" style="width: 100% ;
        height: 400px"></div>
      </div>
    </div>
  </div>
  </body>
</html>
```

So now, go back and carefully look through the very first block of code we showed in this section that was the entire shiny app and realize that code creates all of the preceding HTML and JavaScript. Now, in this example, our R code was not particularly impressive. All the same, it is the principle that counts. Users can now interact with data, and it is not impossible to imagine more interesting applications.

In the next few sections, we introduce a handful of other interesting and useful Shiny input functions along with some essential reactive functions. Finally, in the end, we upload all of them to our cloud so you can share them live on the Web.

13.2 Shiny in Motion

Now that we have a better appreciation for what Shiny does, it is important to consider the dynamic potential of a web page over print reports. Users might engage hands-on with data, allowing research to be more intuitively understood by a wider audience. Information can "move" and shift with time and loop over an animation. With the right underlying data, custom reports and "views" can be user generated via variable selection. This can reduce work on the data science side and give the information consumer precisely what they need.

Another feature of Shiny we mentioned earlier is that R code makes sense to place in three distinct locations. The first is at the beginning of all the code, before user interface or server-side code is written. Code placed in this area runs once per instance. While the value of this may not be as clear on a local machine, when your code is hosted on a cloud, a particular R session may have more than one user interacting with it. That is the way the shiny server we installed in the preceding chapter works. This is a great place to read in any static data or perform any once-only calculations. It is the most computationally efficient. While we do not do so in this example, the code may also be placed inside the server code but outside any reactive functions that would be run once per user connection to a session. Finally, the code inside reactive functions, inside the server function, is executed once every time any input data that the function references is changed.

We take a look at how to change data with respect to time. From the user interface side, there is an input function that is a number-line slider bar. As with all input functions, we need to give it a name so our reactive functions can know where to look for input updates. As with our numericInput() from the prior section, we give this function some minimum and maximum values in years. Finally, we are going to animate our sliderInput() function. Take a look at the code we use inside our user interface. Our code uses an animate option that includes an interval in milliseconds and also has the animation keep on looping:

```
sliderInput("ayear", "Academic Year:",
                 min = 2014, max = 2017,
                 value = 2014, step = 1,
                 animate = animationOptions(interval=600, loop=T) )
```

While this application also shows how to perform R calculations on both static data and user inputs, it really is quite simple again from the R code perspective. In fact, other than the titlePanel() command and of course our slider code, this is quite a bit like our first Shiny application. We show both the code and the end result in Figure 13-4.

```
library(shiny)
slider_data<-read.csv("slider_data.csv", header = TRUE, sep = ",")
Phase1    <- slider_data[,2]
Phase2    <- slider_data[,3]
Phase3    <- slider_data[,4]
Phase4    <- slider_data[,5]

# Define UI for application that draws Bar Plot
ui <- shinyUI(fluidPage(

   # Application title
   titlePanel("Training Programme Results"),

   # Sidebar with a slider input for number of bins
   sidebarLayout(
       sidebarPanel(
sliderInput("ayear", "Academic Year:",
                     min = 2014, max = 2017,
                     value = 2014, step = 1,
                     animate = animationOptions(interval=600, loop=T) ) ),

       # Show the bar plot
       mainPanel(plotOutput("barPlot") ) ) ))

# Define server logic required to draw a barplot
server <- shinyServer(function(input, output) {

   output$barPlot <- renderPlot({
       # Count values in each phase that match the correct date.
       cap<-input$ayear*100
       x <- c(sum(Phase1<cap), sum(Phase2<cap), sum(Phase3<cap),
       sum(Phase4<cap))

       # draw the barplot for the correct year.
```

```
    barplot(x, names.arg = c("Phase I", "Phase II", "Phase III", "Fellows"),
            col = c("deeppink1","deeppink2","deeppink3","deeppink4"),
            ylim=c(0,50))
    })              })
# Run the application
shinyApp(ui = ui, server = server)
```

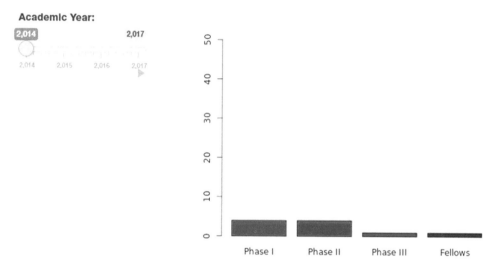

Figure 13-4. *A slider bar input that has a Play button feature showing data in motion*

13.3 Uploading a User File into Shiny

Our last example for this chapter shows how users can upload files via shiny that can then be processed via R. Of course, unlike our first two examples, this is a shade riskier because such uploaded files are not under our control. The data might be any file and not truly work with our results. It is possible to build various error-checking features into code; in fact, we mentioned such techniques in part while discussing functions. However, to focus our code on strictly new features, we pretend for a moment that we live in a perfect world. We have a sample file on the Apress website for this book that is just a comma-separated values file with the numbers 1–10.

Perhaps unsurprisingly, to receive a file input, the command in shiny is
`fileInput()`. Other than `inputId`, maybe the most interesting feature is that this could
even accept multiple uploads, although in our example we leave that to its default value
of `FALSE`. Uploading more than one file might not be supported in all browsers. Also in
the user interface area, we use a new output function, `tableOutput()`, which along with
`plotOutput()` allows us to have both chart and graphical outputs for our users. Before
we talk about the server-side code that enables us to interact with our user's file, we take
a look at what our CSV file upload gives us. Figure 13-5 is the view after we uploaded our
CSV file, and we show the code for just the user interface:

```
ui <- shinyUI(fluidPage(

        fileInput("file", label = h3("Histogram Data File input")),
        tableOutput("tabledf"),
        tableOutput("tabledf2"),
        plotOutput("histPlot1")  ))
```

First of all, notice the first table. It contains four variables that are the results of
directly accessing the `input$file` variable. This might be counterintuitive based
on our other examples, where the user inputs could be directly accessed from their
`input$inputId`. However, it does show our user input is, in fact, a data frame and the
temporary data path is available to access just as any file location might be. Fair warning:
If a user uploads another file, since we are in a temp directory, our old file might well be
overwritten. Next, notice our file's actual data is in the second table, and as shown in the
histogram, that information is available for any R computations we might choose to code.

We next focus on the server side, because this is not as simple as our prior examples.
Sometimes, you have input that takes a little more managing than a single numeric or
text value. Because that data is input data, it needs a reactive function to be alerted if
it has changed. However, you might prefer to use fairly clean code. Fortunately, shiny
offers a basic reactive function fittingly named, `reactive()`. In the following code,
we create our own reactive function to cope with a new scenario, namely, the one we
wanted. When a file is uploaded, shiny alerts our reactive function `reactive()` that
`input$file` has changed. The function naturally queries the input to see what the input
is now and stores that data frame into the variable `file1`. We would like access to the
data inside our CSV file, so we need to call `read.csv()` on that file, which needs to know
the file's location. That is stored in the last column, named `datapath`. What we get is an
object that we can treat like a data frame later in our code:

```
histData<-reactive({
  file1<-input$file
  read.csv(file1$datapath,header=TRUE ,sep = ",")     })
```

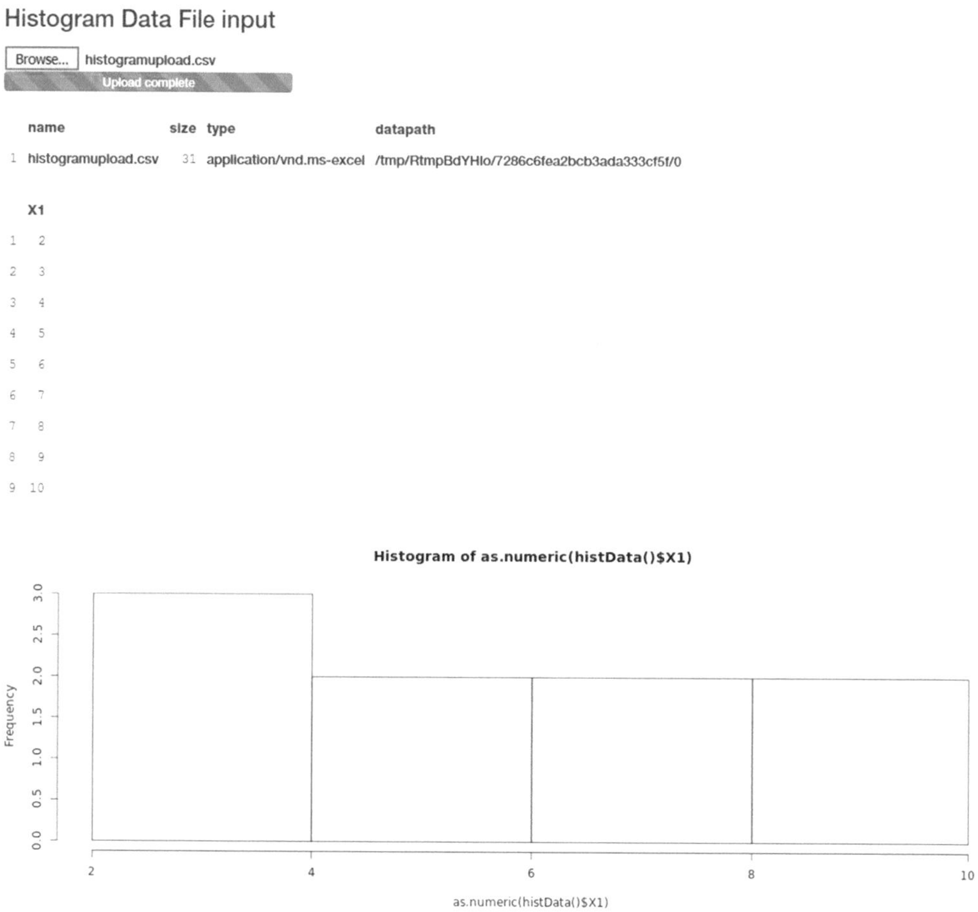
Histogram Data File input
Browse...
histogramupload.csv
Upload complete
name
size
type
datapath
1
histogramupload.csv
31
application/vnd.ms-excel
/tmp/RtmpBdYHlo/7286c6fea2bcb3ada333cf5f/0
X1
1 2
2 3
3 4
4 5
5 6
6 7
7 8
8 9
9 10
Histogram of as.numeric(histData()$X1)
Frequency
0.0
0.5
1.0
1.5
2.0
2.5
3.0
2
4
6
8
10
as.numeric(histData()$X1)

***Figure 13-5.** File upload results*

Other than this, we can use our data now inside other reactive functions, familiar ones even. We show one more new reactive function, `renderTable()`, as well as a familiar one, `renderPlot()`. You should also draw your attention to `histData()$X1`. Notice this has function parentheses, yet elements are still accessed via the familiar $. This is built with a reactive function, and as such, we must acknowledge that it is both a function that takes input from the user and a data storage object with data to access:

```
output$tabledf2<-renderTable({
  histData()                   })
```

```
output$histPlot1<-renderPlot({
  hist(as.numeric(histData()$X1))            })
```

That wraps up our analysis of the code. For completeness, we end this section
with the entire code, but now you have the techniques to be up and running in shiny.
New functions and features are added to this interface quite regularly. However, the
framework is consistent. Thus, it helps to view current vignettes, look for examples and
templates, and bring new visuals into your applications as needed. The next section
of this chapter shows how to upload these application files to the cloud and get these
working live in the cloud. Finally, notice the h3() function inside the fileInput()
function in the user interface. There is a whole collection of functions duplicating
various common HTML features such as headers. We will use these in the following
chapter as we explore ways to make our user interface, for lack of a better word, pretty:

```
library(shiny)

ui <- shinyUI(fluidPage(

    # Copy the line below to make a file upload manager
    fileInput("file", label = h3("Histogram Data File input"), multiple =
    FALSE),
    tableOutput("tabledf"),
    tableOutput("tabledf2"),
    plotOutput("histPlot1")
))

# Define server logic required to draw a
server <- shinyServer(function(input, output) {

   output$tabledf<-renderTable({
       input$file
   })

histData<-reactive({
   file1<-input$file
   read.csv(file1$datapath, header=TRUE ,sep = ",") })
```

```
output$tabledf2<-renderTable({
   histData()              })

output$histPlot1<-renderPlot({
   hist(as.numeric(histData()$X1))                })

})

# Run the application
shinyApp(ui = ui, server = server)
```

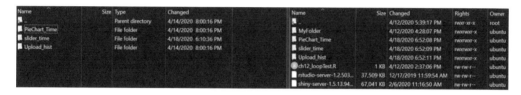

Figure 13-6. *WinSCP with* `PieChart_Time`*,* `slider_time`*, and* `Upload_hist`
applications all uploaded

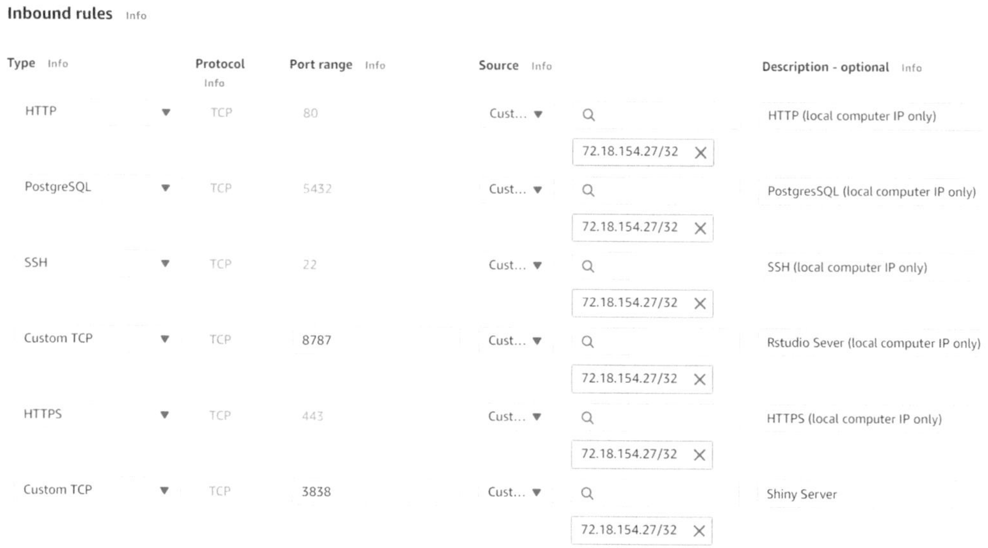

Inbound rules Info

Type Info		Protocol Info	Port range Info	Source Info		Description - optional Info
HTTP	▼	TCP	80	Cust... ▼	Q 72.18.154.27/32 ✕	HTTP (local computer IP only)
PostgreSQL	▼	TCP	5432	Cust... ▼	Q 72.18.154.27/32 ✕	PostgresSQL (local computer IP only)
SSH	▼	TCP	22	Cust... ▼	Q 72.18.154.27/32 ✕	SSH (local computer IP only)
Custom TCP	▼	TCP	8787	Cust... ▼	Q 72.18.154.27/32 ✕	Rstudio Sever (local computer IP only)
HTTPS	▼	TCP	443	Cust... ▼	Q 72.18.154.27/32 ✕	HTTPS (local computer IP only)
Custom TCP	▼	TCP	3838	Cust... ▼	Q 72.18.154.27/32 ✕	Shiny Server

Figure 13-7. *The AWS Inbound table with Shiny Server port 3838 open to one
single local IP address*

Notice in this code, we did not focus on distinctions such as side panels or main panels; instead, we had our user interface side (both the requested input and the generated output) all in one hodgepodge. It is worth noting that if you have a side panel, a main panel is needed to hold any output that is not hosted by the side panel.

13.4 Hosting Shiny in the Cloud

Our last task is to get our files from a local machine to the cloud instance. Remember, in previous chapters, we have used not only PuTTY to install Shiny Server on our cloud but also WinSCP to move files up to our cloud. We start WinSCP, reconnect, and ensure that on the left, local side we are in our `shiny` application folder, while on our right, cloud server side, we are in /home/ubuntu. We show the result of that process in Figure 13-6.

From here, we access PuTTY and run the following code from the command line, one at a time:

```
$ sudo cp -R ~/PieChart_Time /srv/shiny-server/
$ sudo cp -R ~/slider_time /srv/shiny-server/
$ sudo cp -R ~/Upload_hist /srv/shiny-server/
```

This leaves us with one final step. Recall, back when we first set up our cloud instance, we carefully allowed just a few ports to work for just a few IP addresses. At the time we added an entry for Shiny server. It is possible your local machine changed IP addresses. Thus, if the prior steps did not work, remember to get into the AWS *EC2 Dashboard* and update your *security groups* (specifically your *AdvancedR group*). The result looks like Figure 13-7.

You may now verify your apps work by typing the following URLs into any browser:

- `http://Your.Instance.IP.Address:3838/PieChart_Time/`

- `http://Your.Instance.IP.Address:3838/slider_time/`

- `http://Your.Instance.IP.Address:3838/Upload_hist/`

Should you wish to allow others or the world to access your applications, change the *Inbound* source IP address to `0.0.0.0/0` **only** for port 3838. It is a security measure **not** to allow the world access to your *SSH* port. Of course, if you are now hosting an open-to-the-world port 3838, your server's location is visible and known. Thus, it is not a tough guess to imagine that there is an *SSH* port at that address.

If you want the convenience of hosting public, engaging dashboards and also using AWS for large compute operations, one possibility is to have two different cloud servers on two separate security groups. One can be a comparatively pricey instance that has many processors and much RAM. You can be very restrictive on your security settings and even shut down that instance when you are not computing on it to save money. On the other hand, for only serving shiny apps to the public, that can live on an lower-powered machine that is always on and has no confidential data on its hard drive.

13.5 Summary

Using the shiny ecosystem allows R analytics and visuals to be engagingly shared with users with a low tech threshold of entry. Provided your users and consumers are familiar with a web browser, they can be given a chance to explore your R research in ways beyond static PDFs. The shiny ecosystem is constantly improving, so checking online for recent vignettes may provide you just the new functionality needed for improving your ability to effectively communicate your data. That said, the underlying framework of Shiny (as shown in this chapter) has remained fairly static since the first edition of this text. Thus, it is easy to incrementally add on and improve your shiny applications as needed. In Table 13-1, we show some key functions described in this chapter.

Table 13-1. *Listing of key functions described in this chapter and summary of what they do*

Function	What It Does
hiny app.R	A single file template for a Shiny server.
titlePanel()	Sets the HTML page title.
sidebarLayout()	Starts a sidebar.
sidebarPanel	Actual sidebar panel.
mainPanel()	Contains the main interface for the user (often to display visuals).
shinyApp()	Runs the actual Shiny app from R when called.
numericInput()	Collects numeric input from user and passes to server side.
textInput()	Collects text input from user and passes to server side.

(continued)

Table 13-1. (*continued*)

Function	What It Does
checkboxInput()	Collects a check box input from user and passes to server side.
sliderInput()	Collects a slider scale input from user and passes to server side.
fileInput()	Collects an uploaded file input from user and passes to server side.
plotOutput()	User-side displays for a plot.
renderPlot	Server-side reactive creator for a plot.
tableOutput	User-side displays for a table.
renderTable()	Server-side reactive creator for a table.
reactive()	Server-side reactive creator for a generic operation.

CHAPTER 14

Shiny Dashboard Sampler

This chapter is not required in order to understand other chapters in this book. While we introduce some new techniques, what you primarily find here is one entire dashboard sample ready to be modified to suit your needs.

Just what does this sampler do? It takes the applications we used in the preceding chapter and presents them in an engaging, interactive format. It is important to keep in mind that while we often present these applications with a graphical output, any R process can be done to these data. The goal is to allow information consumers the capability to naturally interact with live data, whether those consumers are research reviewers, next-level directors, or board members.

Similarly to the last chapter, we introduce this on Windows inside RStudio [20]. We provide both a framework to understand this dashboard and some pro tips. Please see this sampler as a creativity incubator of sorts rather than a definitive guide. It is our hope you see something you like and, more vitally, realize how you might best showcase your data.

We start with the underlying structure of a dashboard:

```
options(width = 70, digits = 2)
library(knitr)
library(shiny)
library(shinydashboard)
```

14.1 A Dashboard's Bones

Recall our littlest Shiny [6] application from the preceding chapter:

```
library(shiny)

# Define user interface
ui <- shinyUI(fluidPage())
```

M. Wiley and J. F. Wiley, *Advanced R 4 Data Programming and the Cloud*,
https://doi.org/10.1007/978-1-4842-5973-3_14

```
# Define server
server <- shinyServer(function(input, output) {})

# Run the application
shinyApp(ui = ui, server = server)
```

This code was good to look at from the beginning, because it highlights how all Shiny applications comprise a user interface and a server side where the R code logic lives—and these are both run as an R process themselves (possibly on a cloud server). All we intend to add to this structure is some code on the user interface side. This should make sense, because a dashboard is more about user interface than any new code logic. The goal is to make analytics and statistical inferences readily available to users. There is a new library to install for this, but it is just one, `shinydashboard` [5]. From the command line of RStudio, go ahead and run the following code:

```
install.packages("shinydashboard")
```

With this new package installed, we may now take a look at the littlest Shiny dashboard application. Dashboards include a header, a sidebar with various menu items, and the main body in which the different objects of the applications exist. Compare and contrast this dashboard code layout to the single application code. The two versions are almost the same. After the code, see what they generate in Figure 14-1.

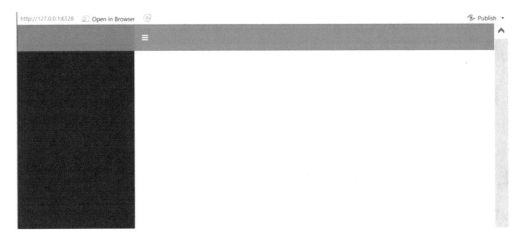

Figure 14-1. *The littlest Shiny dashboard is quite empty*

```
library(shiny)
library(shinydashboard)

# Define user interface
ui <- dashboardPage(
                dashboardHeader(),
                dashboardSidebar(),
                dashboardBody()
                )

# Define server
server <- shinyServer(function(input, output) {})

# Run the application
shinyApp(ui = ui, server = server)
```

As you can see in the preceding code, we have three functions: dashboardHeader(),
dashboardSidebar(), and dashboardBody(). These provide a nice structure for our
dashboard. These all reside inside dashboardPage() as the first three of five formal
arguments for the page function. The final two formals include title (which defaults to
NULL) as well as skin (which takes on one of blue, black, purple, green, red, or yellow
color). As you can see for yourself with the formals(dashboardPage) command, the
page function is indeed simple enough. The title argument controls the heading in the
browser itself (or on the browser tab, if multiple tabs are open in most popular browsers).
The skin color options control the overall color palette of the dashboard. Now, the first
three formals are functions in their own right and become quite complex. Nevertheless,
we consider all three in turn:

```
formals(dashboardPage)

## $header
##
##
## $sidebar
##
##
## $body
##
```

```
##
## $title
## NULL
##
## $skin
## c("blue", "black", "purple", "green", "red", "yellow")
```

Dashboard Header

The dashboardHeader() function also takes a title formal that adds user-defined text into the header area as well as a titleWidth argument controlling the width of that text. The header also accepts several other Shiny user interface–style functions to control page navigation along with a handful of more specialized functions that allow for more interactive navigation. In this chapter, and indeed this book, we essentially ignore these functions to promote shiny and shinydashboard as ways of serving already familiar R code to data consumers rather than learning too many new features all at once. In fact, the last formal we mention for this function is disable, which defaults to FALSE and on TRUE hides the header entirely. In our sampler, we keep the header, yet make few edits to it:

```
formals(dashboardHeader)
```

```
## $...
##
##
## $title
## NULL
##
## $titleWidth
## NULL
##
## $disable
## [1] FALSE
##
## $.list
## NULL
```

Dashboard Sidebar

Moving on to the dashboardSidebar() function, the only explicitly called-out formals are again a disable option along with a width argument. However, there are more functions possible here beyond these mentioned. In general, *any* of the user interface functions from Shiny works in the sidebar. Thus, if it makes sense to place a slider bar or drop-down input functions in your menu, that is possible. There are, however, some functions unique to the dashboard library that make sense to place inside the sidebar area.

The first of these is the sidebarMenu() function. Much like dashboardSidebar() itself, this function primarily takes other inputs and handles the organization of them in an efficient fashion. Of particular note is the menuItem() function, which is called from inside sidebarMenu(). In this sampler, we house each of the applications created in the previous chapter in their own call to menuItem(). As you can see in the following code, this function has several formals:

```
formals(menuItem)

## $text
##
##
## $...
##
##
## $icon
## NULL
##
## $badgeLabel
## NULL
##
## $badgeColor
## [1] "green"
##
## $tabName
## NULL
##
```

```
## $href
## NULL
##
## $newtab
## [1] TRUE
##
## $selected
## NULL
##
## $expandedName
## as.character(gsub("[[:space:]]", "", text))
##
## $startExpanded
## [1] FALSE
```

In the preceding function and formal arguments, the text takes a text string in quotes that provides the visible link name for that particular item on the menu. The icon formal is given a name for a *Font Awesome* icon. We provide a link to their online library, with all possible icon choices, in the summary section of this chapter. There are other icon libraries possible; we save such explorations for you. The badgeLabel call puts an inline badge on the far right of the sidebar with a short bit of text that you choose. Badges do a good job of calling your viewers' attention to a particular menu item and, consequently, to the underlying application. Many of these formals are somewhat optional; however, there must be either a tabName linked to a related tabItem() in the dashboard's body or there must be a URL given to the href command. If a URL is given rather than a tabName, then newtab can be changed from a default of TRUE to FALSE should you wish your viewer to leave your site when visiting that page. Finally, it is possible to force the default to a particular tab, away from the first menuItem(), by setting selected to TRUE. Be sure to only do this once for your dashboard!

As we turn our attention to the main body of our dashboard, we remind you the sidebar can also receive any Shiny function—such as the img() function which points to an image that must be contained in a folder named www inside your dashboard application's folder. We show a code snippet (which should not be run yet) of our sampler's menu along with the result in Figure 14-2. The names of the first three menu items should be familiar from the previous chapter:

```
sidebarMenu(
  menuItem(
    "Pie Charts!",
    tabName = "PieChart_Time",
    icon = icon("pie-chart")
  ),
  menuItem(
    "Use a Slider",
    tabName = "slider_time",
    icon = icon("sliders"),
    badgeLabel = "New"
  ),
  menuItem(
    "Upload a Histogram File",
    tabName = "Upload_hist",
    icon = icon("bar-chart"),
    badgeLabel = "Live"
  )
)
```

Figure 14-2. *The result of some sidebarMenu() and menuItem() functions*

Figure 14-3. *A dashboard with a* menuItem() *in the sidebar and a* tabItem() *in the body. Notice the matching* First

Dashboard Body

As you saw in the preceding code, each menu item has a tabName. We now discuss the code inside the dashboardBody() function that catches tabName. Take a look at the little, although not the littlest, Shiny dashboard in the following code and Figure 14-3, and then we will walk through the new lines:

```
library(shiny)
library(shinydashboard)

# Define user interface
ui <- dashboardPage(
    dashboardHeader(),
    dashboardSidebar(sidebarMenu(menuItem("DemoLink", tabName = "First"))),
    dashboardBody(tabItems(tabItem(tabName = "First", "Main area or 'body'
    text."))),
    skin = "yellow")

# Define server
server <- shinyServer(function(input, output) {})

# Run the application
shinyApp(ui = ui, server = server)
```

Many items might be placed inside the body of our dashboard. However, if we want to use the tabs that we coded back in the sidebar section, we do a general call to tabItems(). This function takes no arguments except individual tabItem() function calls; each tabItem() needs to have a unique tabName **matching** a unique tabName from the sidebarMenu().

What goes inside each tabItem()? Well, just as in Shiny, there is a benefit to making the layout fluid enough that your dashboard can cope with various-sized browser windows (e.g., mobile vs. desktop platforms). Thus, the common function fluidRow() makes a reappearance. Recall fluidRow() is set up to have a row width of at most 12 units. Remembering that fact, keep a sharp lookout at the default widths for each of the three box objects we discuss to fill those rows. Now, there are more than these three objects; again, our goal is to get you quickly to the point where you may serve your data via the cloud to your users or stakeholders.

The most basic input element to go in a row is box(). Inside this function, we place any of the application code we used in the preceding chapter to either solicit input from users or to output reactive content. However, box() also takes nine specific arguments. Naturally, each box() may take a title value, a user-defined text string that should succinctly describe the contents of the box or the action required. At the bottom of the box, there is a footer, which may also take a text string. Boxes also have a status that may be set to take on values of primary, success, info, warning, or danger. Each status value has an associated color that is part of the default Cascading Style Sheets structure that makes shinydashboard look rather pretty. The colors are fairly consistent, although they may have some variance depending on which color you selected for the whole dashboard. The default value for solidHeader is FALSE, although this, of course, may be changed to TRUE. The positive choice simply enhances the color of the status to become a narrow banner background to the title text. Should you wish the main area of your box to have a specific color, background can be set from approximately 15 valid color choices. Input and Output code displays will visually appear on top of that background, as the name suggests. The default width for a box is 6, so two boxes fill up a row unless you take steps to narrow them. Failure to keep the total fluidRow() width of 12 in mind can lead to some odd layouts. Box height can also be set to a fixed height. This can be helpful if you require very even rows. Finally, boxes are collapsible and may be already collapsed, although both options default to FALSE. Figure 14-4 shows some results of these various settings:

```
formals(box)

## $...
##
##
## $title
```

```
## NULL
##
## $footer
## NULL
##
## $status
## NULL
##
## $solidHeader
## [1] FALSE
##
## $background
## NULL
##
## $width
## [1] 6
##
## $height
## NULL
##
## $collapsible
## [1] FALSE
##
## $collapsed
## [1] FALSE
```

Another type of box is valueBox(), which takes six formal arguments. The first is value, which is usually a short text string or value, as shown in Figure 14-5. The valueBox() also takes a subtitle argument, which is again a text string. You can choose an icon, although the default value is NULL. These boxes come with a default color of aqua and are rather short with a default width of only 4. Finally, valueBox() also takes a value for href, which defaults to NULL. We show the code we used in our sampler's value box here:

```
valueBox(
    endYear - startYear,
    subtitle = "Years of Data",
```

```
    icon = icon("band-aid"),
    width = 2
)
```

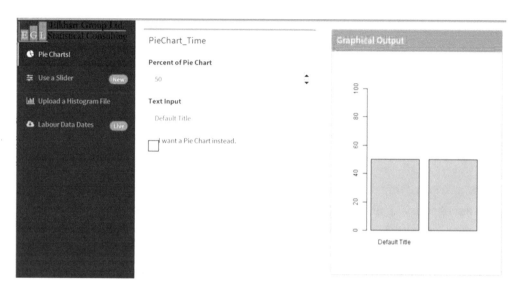

Figure 14-4. *A familiar application in two boxes with various options—see section "Complete Sampler Code" for details*

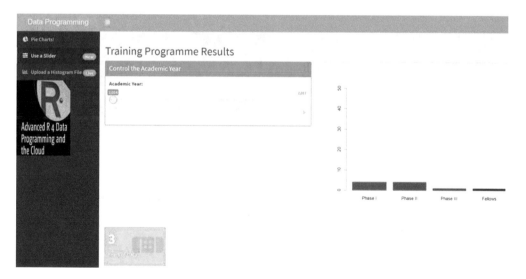

Figure 14-5. *Sample* valueBox() *showing the number of days of data that were pulled live from a website*

We finish out our discussion of the dashboardBody() boxes with a box similar to the valueBox(): the infoBox(). This box comes with a title that takes a text string, a value to highlight and display in the box, optional subtitle text, and a default icon of a "bar chart". It also features defaults for color = aqua, width = 4, and href = NULL. What is different is the icon is smaller and there is a fill option, which is usually FALSE, that can change the way the background color displays. We do not show this box, but it does look quite similar to the small box in Figure 14-5. In our experience, these boxes are not significantly different other than in their appearance.

14.2 Dashboard in the Cloud

We do want to serve this dashboard into our cloud instance, and this is a slightly more complex set of code. Again, we created this in a standard Shiny folder by using RStudio, so it comes in its own file. The file structure on Windows looks like Figure 14-6. There are three files in our shinydashboard_cloud folder. Note the www subfolder, which hosts our sampler's image, followed by the app.R file, which holds our entire Shiny application as well as our CSV file.

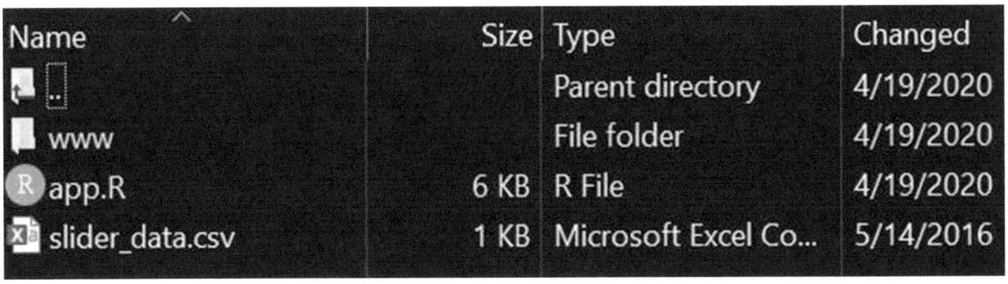

Name		Size	Type	Changed
↰ [..]			Parent directory	4/19/2020
📁 www			File folder	4/19/2020
ⓡ app.R		6 KB	R File	4/19/2020
📄 slider_data.csv		1 KB	Microsoft Excel Co...	5/14/2016

Figure 14-6. *The file structure for shinydashboard_cloud*

We use WinSCP to upload our files and folders to our Ubuntu cloud instance, as shown in Figure 14-7. Remember, while the full code is shown in the following section, the complete set of code is available online at the Apress site and/or our GitHub location for this book.

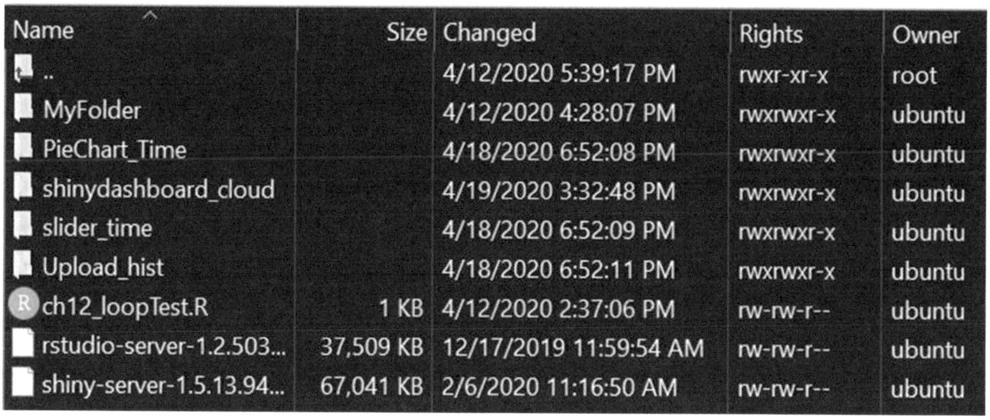

Name	Size	Changed	Rights	Owner
..		4/12/2020 5:39:17 PM	rwxr-xr-x	root
MyFolder		4/12/2020 4:28:07 PM	rwxrwxr-x	ubuntu
PieChart_Time		4/18/2020 6:52:08 PM	rwxrwxr-x	ubuntu
shinydashboard_cloud		4/19/2020 3:32:48 PM	rwxrwxr-x	ubuntu
slider_time		4/18/2020 6:52:09 PM	rwxrwxr-x	ubuntu
Upload_hist		4/18/2020 6:52:11 PM	rwxrwxr-x	ubuntu
ch12_loopTest.R	1 KB	4/12/2020 2:37:06 PM	rw-rw-r--	ubuntu
rstudio-server-1.2.503...	37,509 KB	12/17/2019 11:59:54 AM	rw-rw-r--	ubuntu
shiny-server-1.5.13.94...	67,041 KB	2/6/2020 11:16:50 AM	rw-rw-r--	ubuntu

Figure 14-7. *WinSCP upload folder view*

Logging into PuTTY, we do have one bit of code to run. Our sampler files need to be moved from our user's folder to where the Shiny server expects them to live:

```
$ sudo cp -R ~/shinydashboard_cloud /srv/shiny-server/
```

This completes our work with Shiny and the Shiny dashboard. Your dashboard should be live at `http://Your.Instance.IP.Address:3838/shinydashboard_cloud/` and viewable from your web browser. We have a few more comments to make before we close out the chapter.

First, we again suggest you visit the Apress site for this textbook and download the code bundle for this chapter. It contains the complete, working files and folders shown in Figure 14-6 and uploaded in Figure 14-7. There is no need to manually type all this code to see whether a dashboard might be something you would like to have.

Second, we remind you that if you followed our security-minded advice about inbound policy settings on the AWS servers, you might need to change those settings to allow more people than just yourself to view your final dashboard project. Remember, the setting you need to change on the inbound side is to allow port 3838 to be accessed from `0.0.0.0/0`, which is the entire Internet. Keep in mind the entire Internet would then be able to access your dashboard. If you have sensitive data or even information that just does not need to be out and about, please consider partnering with an IT professional or at least an IT power user.

Finally, we hope you enjoyed this chapter and the results as much as we have. Be sure to explore the links in the "Summary" section. You will find valuable information ranging from additional functions, a whole list of icons, a whole second icon library, and the very latest in shiny and shinydashboard package information. These packages are updating fairly regularly, and new features can often provide awesome results.

14.3 Complete Sampler Code

As you review the code that follows, remember to compare and contrast the code with what you learned earlier in this chapter. See the small pieces working together to create a whole. Also, while we did keep this in a single app.R file for this example, the global, server, and user interface pieces of shiny can be split into three files (indeed it is one of the options in RStudio when creating a new *Shiny web app*). This can be quite helpful in segmenting larger dashboards such as the one(s) you may create:

```
library(shiny)
library(shinydashboard)

#run-once code
slider_data <- read.csv("slider_data.csv", header = TRUE, sep = ",")
Phase1    <- slider_data[, 2]
Phase2    <- slider_data[, 3]
Phase3    <- slider_data[, 4]
Phase4    <- slider_data[, 5]

startYear <- 2013
endYear <- 2016

# Define UI for dashboard
ui <- dashboardPage(
  dashboardHeader(title = "Data Programming and the Cloud"),
  dashboardSidebar(
    sidebarMenu(
      menuItem(
        "Pie Charts!",
        tabName = "PieChart_Time",
        icon = icon("pie-chart")
      ),
```

```
    menuItem(
      "Use a Slider",
      tabName = "slider_time",
      icon = icon("sliders"),
      badgeLabel = "New"
    ),
    menuItem(
      "Upload a Histogram File",
      tabName = "Upload_hist",
      icon = icon("bar-chart"),
      badgeLabel = "Live"
    )
  ),
  img(
    src = "R4_DataProgandCloud.JPG",
    height = 200,
    width = 150
  )
),
dashboardBody(tabItems(
  # PieChart_Time Content
  tabItem(tabName = "PieChart_Time",
          fluidRow(
            box(
              title = "PieChart_Time",
              status = "warning",
              numericInput(
                "pie",
                "Percent of Pie Chart",
                min = 0,
                max = 100,
                value = 50
              ),

              textInput(
                "pietext",
                "Text Input",
```

```
                  value = "Default Title",
                  placeholder = "Enter Your Title Here"
                ),

                checkboxInput("pieChoice",
                              "  I want a Pie Chart instead.",
                              value = FALSE)
              ),

              box(
                title = "Graphical Output",
                solidHeader = TRUE,
                status = "warning",
                plotOutput("piePlot")
              )
            )),

    # Slider Tab Content
    tabItem(
      tabName = "slider_time",
      h2("Training Programme Results"),
      fluidRow(
        box(
          title = "Control the Academic Year",
          status = "primary",
          solidHeader = TRUE,
          sliderInput(
            "ayear",
            "Academic Year:",
            min = 2014,
            max = 2017,
            value = 2014,
            step = 1,
            animate = animationOptions(interval = 600, loop = T)
          )
        ),
```

```
      box(plotOutput("barPlot"))
    ),
    fluidRow(
      valueBox(
        endYear - startYear,
        "Years of Data",
        icon = icon("band-aid"),
        width = 2
      )
    )
  ),

  #Histogram from an Uploaded CSV
  tabItem(tabName = "Upload_hist",

          fluidRow(
            box(title = "File Input",
    # Copy the line below to make a file upload manager
                fileInput(
                  "file",
                  label = h3("Histogram Data File input"),
                  multiple = FALSE
                )),
            box(
              title = "Data from file input",
              collapsible = TRUE,
              tableOutput("tabledf")
            )
          ),

          fluidRow(
            box(tableOutput("tabledf2")),
            box(background = "blue",
                plotOutput("histPlot1"))
          ))

)),
```

```
  title = "Dashboard Sampler",
  skin = "yellow"

)

################### Define server logic required to draw a histogram
server <- shinyServer(function(input, output) {
  output$piePlot <- renderPlot({
    # generate Pie chart ratios based on input$pie from user
    y <- c(input$pie, 100 - input$pie)

    # draw the pie chart or barplot with the specified ratio and label

    if (input$pieChoice == FALSE) {
      barplot(y,
              ylim = c(0, 100),
              names.arg = c(input$pietext, paste0("Complement of ",
                                                  input$pietext)))

    } else{
      pie(y, labels = c(input$pietext, paste0("Complement of ",
                                              input$pietext)))

    }

  })

  output$barPlot <- renderPlot({
    # Count values in each phase which match the correct date.
    cap <- input$ayear * 100
    x <- c(sum(Phase1 < cap),
           sum(Phase2 < cap),
           sum(Phase3 < cap),
           sum(Phase4 < cap))

    # draw the barplot for the correct year.
    barplot(
      x,
      names.arg = c("Phase I", "Phase II", "Phase III", "Fellows"),
```

```
      col = c("deeppink1", "deeppink2", "deeppink3", "deeppink4"),
      ylim = c(0, 50)
    )
})

####Here is where the input of a file happens

output$tabledf <- renderTable({
  input$file
})

histData <- reactive({
  file1 <- input$file
  read.csv(file1$datapath, header = TRUE , sep = ",")
})

output$tabledf2 <- renderTable({
  histData()
})

output$histPlot1 <- renderPlot({
  hist(as.numeric(histData()$X1))
})

###############end of server####################
})

# Run the application
shinyApp(ui = ui, server = server)
```

14.4 Summary

References and more are shown in Table 14-1.

Table 14-1. *Listing of key functions described in this chapter and summary of what they do*

Function	What It Does
`menuItem()`	A "link" in your dashboard - `tabName = ""` must match a `tabItem()`.
`tabName = ""`	A common pair identifier in menu and tab items.
`icon = ""`	Uses one of the following icon links to find the text name for your icon.
`badgeLabel = ""`	Takes a text string you write and places it beside a "link." Not much space.
`tabItem()`	A "page" in your dashboard - `tabName = ""` must match a `menuItem()`.
`fluidRow()`	Rows are of length 12 "units." A single `tabItem()` may have several rows.
`box()`	A wrapper that holds some reactive code (e.g., a graph or a place for user input).
`valueBox()`	A good way to call out some specific info—it can be reactive (or not).
shinydashboard homepage	`https://rstudio.github.io/shinydashboard/`
Icons	`http://fontawesome.io/icons/`
Other icons	`https://icons.getbootstrap.com/`
shiny function list	`http://shiny.rstudio.com/reference/shiny/latest/`
Shiny Server Guide	`http://docs.rstudio.com/shiny-server/`
Shiny Gallery	`http://shiny.rstudio.com/gallery/`
shiny CRAN documentation	`https://cran.r-project.org/web/packages/shiny/shiny.pdf`

CHAPTER 15

Dynamic Reports

In the preceding chapters, you saw how Shiny delivers interactive environments based on dynamic data. Dynamic reports and the live dashboards have many similarities. On the other hand, this chapter, because the reports end up as PDFs, is less interactive. Reports are a fact of life in many fields, and stakeholders tend to require snapshots in time rather than fully interactive environments. Through the knitr and rmarkdown packages, we create documents (e.g., PDF, HTML, or Microsoft Word) based on data input. For regular reports that build on continuously changing data, yet that have the same structure overall, this is a great time-saver.

The `rmarkdown` package [42] adds the capability to embed R processes into such documents, allowing one-click analytics that are beautifully formatted using a variety of styles. The LaTeX interface achieves almost any formatting required, although there are options besides PDF. From boardroom reports on enrolment based on live pulls from a student information system database to strategic plan action-sheets drawn from key performance indicators, dynamic reports work best when variable input needs shaping into standardized output. This is also facilitated via `knitr` [38] and formatR packages [39].

Additionally, in this chapter, we use the `evaluate` package [34] and `tufte` package [41]. These packages allow us to organize the layout and format of our data more elegantly. We also use the `ltm` package [19] for the psychometric tests.

Our goals for this chapter are to enhance our software environment both on our local and cloud machines to allow the dynamic documents to compile, explore the structure of standalone `rmarkdown`, develop a Shiny page that allows a user to upload data and then download a PDF report, and finally push this all to our cloud instance. We start with needed software:

```
options(width = 70, digits = 2)
```

© Matt Wiley and Joshua F. Wiley 2020
M. Wiley and J. F. Wiley, *Advanced R 4 Data Programming and the Cloud*,
https://doi.org/10.1007/978-1-4842-5973-3_15

15.1 Needed Software

We already have a great deal of the software needed, namely, R, the Shiny packages [6], PuTTY, and our cloud instance on Ubuntu. We need a handful of new packages installed—both on the local machine and the cloud server.

Local Machine

First, in your R console, it is important to run the following code independent of any R file:

```
install.packages("rmarkdown")
install.packages("tufte")
install.packages("knitr")
install.packages("evaluate")
install.packages("ltm")
install.packages("formatR")
library(knitr)
```

LaTeX Product

When using R to generate PDFs, it is best to only have a single LaTeX engine on the computer doing the work. As mentioned in a prior chapter, there are many options for LaTeX, including both tinytex (from CRAN) and MiKTeX [21]. If you already have a LaTeX product, it should work with RStudio (on Windows at least, you may need to make sure your LaTeX solution is in the global or user *environment PATH*).

If you do not yet have such a system, MiKTeX is a great option, and its default settings and installation options generally "simply work" on most configurations. If you do reach a point where a PDF is **not** generated, in our experience, searching on the output error message usually leads to some workable suggestions.

To install MiKTeX to create PDF files, visit the MiKTeX site (http://miktex.org/download) to obtain either the recommended basic installer (which is 32-bit) or the other basic installer (which is 64-bit). We use the 64-bit version for our systems, although in this chapter there is not likely to be a major difference between the versions. After download, accept all the defaults during installation. After installation, there may be updates to the MiKTeX package. To perform these updates, go to your start files.

In a folder titled MiKTeX, you will find an option for *MiKTeX Console*. There is a button titled *Check for updates* that you should click and run. While likely not required, we do recommend a restart after this installation.

Cloud Instance

Using PuTTY, access your cloud instance console. We need to run each line of code that follows individually at the console. Recall from an earlier chapter that these install the packages for all users of that server, not just your user. This is key to allow the Shiny server access to the correct packages:

```
$ sudo su - -c "R -e \"install.packages('knitr')\""
$ sudo su - -c "R -e \"install.packages('rmarkdown')\""
$ sudo su - -c "R -e \"install.packages('tufte')\""
$ sudo su - -c "R -e \"install.packages('ltm')\""
$ sudo su - -c "R -e \"install.packages('formatR')\""
$ sudo su - -c "R -e \"install.packages('data.table')\""
$ sudo su - -c "R -e \"install.packages('evaluate')\""
```

The preceding code does not take very long. However, installing texlive (the Ubuntu solution for LaTeX used instead of MiKTeX) may take some time. It also demands a fair bit of disk space, so be sure you have room. In our trials, the full version is required. On the cloud, as on your local machine, it is good to not have more than one version of LaTeX. Thus, if you are using a cloud instance you use for other things and if you already have tinytex package installed, you may need to experiment to see if you can get away with not using texlive:

```
$ sudo apt-get install texlive-full
$ sudo reboot
```

After this, we recommend a sudo reboot to restart your cloud instance. This is a good place to do that and then close PuTTY for a bit. While your server restarts, you may use your local machine to explore dynamic documents.

15.2 Dynamic Documents

It is not required to link dynamic documents to the Internet in any way. Indeed, most reports may be internal. Inside RStudio, selecting *File*, *New File*, and *R Markdown* opens a wizard. Notice there are options for documents (e.g., HTML, PDF, or Word) as well as presentations (e.g., HTML or PDF). A title is customary, as is listing the author(s). Of course, RStudio is not required; creating a new file in R with the `.Rmd` suffix suffices.

We intend to create a hybrid approach to this introduction to dynamic documents. That is, we first give a straightforward and informative example of what such code looks like, and then we follow up with a more in-depth analysis of each piece of that code. We start with a new file titled `ch15_html.Rmd` and fill it with the following code:

```
---
title: "ch15_html"
author: "Matt & Joshua Wiley"
date: "29 April 2020"
output: html_document
---
"'{r echo=FALSE}
##notice the 'echo=FALSE'
##it means the following code will never make it to our knited html
library(data.table)
diris <- as.data.table(iris)
##it is still run, however, so we do have access to it.
"'

#This is the top-level header; it is not a comment.

This is plain text that is not code.

If we wish *italics* or **bold**, we can easily add those to these
documents.
Of course, we may need to mention 'rmarkdown' is the package used,
and inline code is nice for that.
If mathematics are required, then perhaps x~1~^2^ + x~2~^2^ = 1 is wanted.

##calling out mathematics with a header 2
On the other hand, we may need the mathematics called out explicitly,
in which case $x^{2} + y^{2}= \pi$ is the way to make that happen.
```

###I like strikeouts; I am less **clear** about level 3 headers.
I often **find** when writing reports that I want to say something
is ~~absolutely foolish~~ obviously relevant to key stakeholders.

######If you ever write something that needs Header 6
Then I believe, as this unordered list suggests:

* You need to embrace less order starting now

    - Have you considered other careers?

```
"'{r echo=FALSE}
summary(diris)
"'
```

However, notice that one can include both the code and the console output:

```
"'{r}
hist(diris$Sepal.Length)
"'
```

The results of knitting the preceding code are shown in Figure 15-1.

As you can imagine, it is the summary data or the graph that is easily regenerated based on any input. In fact, one of the authors uses dynamic documents to create case studies (and a case study key) for students in an introductory statistics course. It is easy enough to generate pseudo-random data with different seeds that allow each student to have a unique data set (and several topics). It also allows for solution keys for fast marking.

Now that you have an overview of rmarkdown, let's explore the previous code in sensible blocks. The first block is YAML, which stands for *Yet Another Markup Language* or possibly the recursively defined *YAML Ain't Markup Language* [18] header. At its simplest, it may contain only information about the title, author, date, and output type. Here, for this document, we selected *html_document*. However, as stated previously, there are more options such as *pdf_document, word_document, odt_document, rtf_document*, and *beamer_presentation* (to name a few—there are in fact more). These types create, respectively, HTML pages, PDFs, Microsoft Word files, OpenDocument or rich text, and Beamer PDF slides. Both Beamer and PDFs require the MiKTeX installation we did earlier. To change the type of document compiled, simply change the output! It is worth noting that the first time you use MiKTeX, it may take longer to compile, as there may need to be a package or two installed behind the scenes. This does not take any user intervention; it simply takes time. It is also worth noting that if you use Beamer presentation, then what can and cannot fit on one slide needs to be considered. In that case, a carriage return followed by —-, followed by another

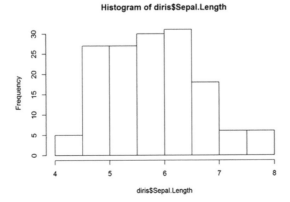

Figure 15-1. *The HTML output of a simple dynamic document*

carriage return, signals new slides in your deck.

```
---
title: "ch15_html"
author: "Matt & Joshua Wiley"
date: "29 April 2020"
output: html_document
---
```

Headers can be more complex than this. R code can be injected inline, using `'r code goes here'` formatting. Thus, we could change the date to be more dynamic by swapping out date "29 April 2020" for date `"'r Sys.Date()'"` in our code. Dozens of render options work with the YAML header, and wiser heads than ours have created templates that use variants of those headers as well as set several nice options. As in Shiny, there is a nearly limitless ability to customize precisely how a document outputs and displays.

The next region of our markdown code is a code chunk. Code chunks start with `"'{r }` and end with `"'`. Between those lives your code. Now, inside the `{r }` there are several useful options. If you have part of your code that is intensive to create and mostly static, you may want to set `cache = TRUE` instead of the default `FALSE`. This stores the results. In our work, we use `echo=FALSE` often, which prevents the code itself, yet not the results of that code, from displaying. Notice that in the first chunk that follows, there is no sign in our final document (Figure 15-1) that the code ran—at least not a sign as far as showing the actual code to the reader. However, the second chunk still shows the results of `summary(diris)`:

```
"'{r echo=FALSE}
library(data.table)
diris <- as.data.table(iris)
"'
```

```
"'{r echo=FALSE}
summary(diris)
"'
```

For code chunks that involve plots or figures, there are many options too. The `fig.align` option may be set to right, center, or left. There are height and width settings for plots with `fig.height` or `fig.width` that default to inches:

```
"'{r fig.align="right"}
hist(diris$Sepal.Length)
"'
```

Finally, there is the text itself, which we do not repeat here because that is of less interest. We do note that code can be written inline as well as just as in the header. This option becomes very useful later, when we want to provide different narrative text depending on our computation results.

15.3 Dynamic Documents and Shiny

We turn our attention now to providing not only a dynamic document but one that others may control via their dynamic input. What we build is a simple Shiny environment that allows for a user to upload a CSV file with quiz score data for students. The data is coded as follows: the columns represent specific questions on the quiz; the rows represent individual students; and the quiz is multiple choice, with 1 coded in for correct responses and 0 coded in for incorrect options.

As in the preceding section, we look at all our code for each of the three files it takes to make this work. Then, we break that code down line by line. For readers already comfortable with Shiny, please feel free to skip to the last portion. We build this code in a folder named `15_DynamicReports`. Later, when we upload to our Ubuntu server, we upload that entire folder.

server.R

On the Shiny server side, we call our libraries and then build the systems needed to upload our CSV file, provide confirmation for users that their file is uploaded, read that file into R, pass the needed information along to our markup document, and then allow our users to download that document as a PDF. It takes just the following 55 lines of code:

#Dynamic Reports and the Cloud

```r
library(shiny)
library(ltm)
library(data.table)
library(rmarkdown)
library(tufte)
library(formatR)
library(knitr)
library(evaluate)

function(input, output) {
  output$contents <- renderTable({
    inFile <- input$file1

    if (is.null(inFile))
      return(NULL)

    read.csv(
      inFile$datapath,
      header = input$header,
      sep = input$sep,
      quote = input$quote
    )
  })

  scores <- reactive({
    inFile <- input$file1
    if (is.null(inFile))
      return(NULL)

    read.csv(
      inFile$datapath,
      header = input$header,
      sep = input$sep,
      quote = input$quote
    )
  })
```

```
output$downloadReport <- downloadHandler(
  filename = function() {
    paste('quiz-report', sep = '.', 'pdf')
  },

  content = function(file) {
    src <- normalizePath('report.Rmd')

    owd <- setwd(tempdir())
    on.exit(setwd(owd))
    file.copy(src, 'report.Rmd', overwrite = TRUE)

    knitr::opts_chunk$set(tidy = FALSE, cache.extra =
    packageVersion('tufte'))
    options(htmltools.dir.version = FALSE)

    out <- render('report.Rmd')
    file.rename(out, file)
  }
 )
}
```

The preceding code started with a call to all our libraries and then quickly got into Shiny's function(input, output) {} server-side wrapper. After that, there are just three main blocks of code. In turn, we look at each of these, starting with the renderTable({}) call. This generates output used in our user interface side. In particular, it is creating a table. It is an interactive function, and when a user uploads a file on the user side, Shiny alerts this function that one of its inputs has changed. This triggers a run of the code. From the side of entry, input$datapath is the location on the server where Shiny stores an uploaded file. Provided there is, in fact, a file, read.csv() can access the file's data path and also gets information about whether the data has a header, what type of separator was used, and some other information set by the user. In fact, much more could be asked of the user that may allow for more customized analytics. Here, we keep the example simple yet efficacious. The result of this read creates a table and stores that table inside the output variable named contents. This has a net effect of Shiny now alerting the user interface side there is possible output. We see where that output goes later in this chapter:

```
output$contents <- renderTable({

    inFile <- input$file1

    if (is.null(inFile))
      return(NULL)

    read.csv(inFile$datapath, header = input$header,
            sep = input$sep, quote = input$quote)
  })
```

In addition to creating a table so our user may immediately see whether their file upload is successful (and we envision in a live scenario perhaps adding comments to the user interface to allow a user to spot a poor upload), we must also read the data in a format that will allow our markdown file to use that data. We read the data into a variable named scores. This is a generic, created-by-us reactive({}) function that is always on the alert for changes in the relevant input variables (in this case, the uploading of a file).

This is one of the reasons Shiny is so customizable. While you already met functions that are reactive, those have all had a targeted purpose (and are named accordingly). This function, on the other hand, is different. It only is reactive, and all behavior is completely determined by what we make it do:

```
scores <- reactive({
    inFile <- input$file1

    if (is.null(inFile))
      return(NULL)

    read.csv(inFile$datapath, header = input$header,
            sep = input$sep, quote = input$quote)
  })
```

The last piece of the server side is the most advanced. This is what controls the download file the user can access. The first part of this block controls the name of the report the user downloads. We named ours quiz-report.pdf, which only looks complex. That code gives us future flexibility should we change the type or nature of the report; it is easy to change our file type or name. Note that if you use RStudio's internal browser, the report downloads with a different name. Next, we call normalizePath() on our markdown file, which returns the file path of the current working directory with the filename in the argument appended. In particular, it returns this based on the operating system environment in which the code exists. This is important for two reasons. First, it allows portability of the code from Windows to Ubuntu (or any operating system). Second, we are about to change our working directory. Thus, a direct call to report.Rmd will no longer work.

It is important we change our working directory; we are about to create a new PDF file. Now, on a local machine we own and have administrative privileges to use, creating a new file is no trouble at all. However, our goal is to upload to the cloud, and we are not assured such privileges on all servers. Thus, we change our working directory to the temporary directory of the server. In R, setwd(tempdir()) does two things. It both copies and returns the current working directory; and then, after that, it changes the directory to the argument. Thus, owd is the *old working directory*. We set an on.exit() parameter so that, once downloadHandler is done, the working directory returns to what it ought to be, so that our Shiny site does not break.

We now copy our markdown file from its original location in the old working directory to our temp area, thereby allowing it to run and generate the PDF in that temp directory. We also set some environmental options for knitr (which creates the markdown PDF) that makes it possible to use the tufte format (named for Edward Tufte, http://rmarkdown.rstudio.com/tufte_handout_format.html). All that remains is to render() our markdown document and rename our file to what we wanted:

```
output$downloadReport <- downloadHandler(
  filename = function() {
    paste('quiz-report', sep = '.', 'pdf')
  },

  content = function(file) {
    src <- normalizePath('report.Rmd')

    owd <- setwd(tempdir())
    on.exit(setwd(owd))
    file.copy(src, 'report.Rmd', overwrite = TRUE)
    knitr::opts_chunk$set(tidy = FALSE, cache.extra =
    packageVersion('tufte'))
    options(htmltools.dir.version = FALSE)
```

```
    out <- render('report.Rmd')
    file.rename(out, file)
}   )
```

What we have seen is the Shiny server side needs to be able to read in the user's uploaded file with the appropriate input information, such as the existence of a header, and then is responsible for starting the rmarkdown process. Admittedly, getting that process to work (along with a Shiny server) on any computer involves changing the working directory as well as some fiddling with layout and formatting options. We conclude this section by noting that a simple PDF would be easier to set up (although we admit to having a preference for the cleaner tufte layout). We turn our attention to the user interface.

ui.R

The user has a fairly clean view for this very simple Shiny application. In just 40 lines of code (and you could remove several options, really), the user can upload the CSV file, view the rendered table from that file, and download the analysis of the file. As before, we first give the entire code and then describe each block in turn:

```
#Dynamic Reports and the Cloud User Interface
shinyUI(fluidPage(title = 'Score Control Panel',
                sidebarLayout(
                    sidebarPanel(
                        helpText(
                            "Upload a wide CSV file where
                each column is 0 or 1 based on whether the student
                got that question incorrect or correct."
                            ),

                        fileInput(
                            'file1',
                            'Choose file to upload',
                            accept = c(
                                'text/csv',
                                'text/comma-separated-values',
                                'text/tab-separated-values',
```

```
                     'text/plain',
                     '.csv',
                     '.tsv'
                   )
                 ),
                 tags$hr(),
                 checkboxInput('header', 'Header', TRUE),
                 radioButtons('sep', 'Separator',
                             c(
                                 Comma = ',',
                                 Semicolon = ';',
                                 Tab = '\t'
                             ),
                             ','),
                 radioButtons(
                   'quote',
                   'Quote',
                   c(
                     None = '',
                     'Double Quote' = '"',
                     'Single Quote' = "'"
                   ),
                   '"'
                 ),
                 tags$hr(),
                 helpText("This is where helpertext goes"),
                 downloadButton('downloadReport')
                   ),
                 mainPanel(tableOutput('contents'))
                   )))
```

As usual for the user interface, we have fluidPage() along with sidebarPanel() and mainPanel(). The main events occur inside the sidebar, and we look there first. The fileInput() function takes our user's input file. We can control the types of files accepted, and we limit ourselves to comma- or tab-separated values. We name the upload file1, recalling on the server side we used input$file1 to read in the file:

```
fileInput('file1', 'Choose file to upload',
          accept = c(
             'text/csv',
             'text/comma-separated-values',
             'text/tab-separated-values',
             'text/plain',
             '.csv',
             '.tsv'
          )
),
```

The only other line of major interest to us is the download area. It is not particularly complex to understand:

```
downloadButton('downloadReport')
```

Although there are helper text boxes to explain to the user, and we do recommend in real life setting up sample layouts so your users understand the acceptable inputs, we opted to keep this user interface as streamlined as possible. Before we turn our attention to the actual report file, we show the Shiny application in Figure 15-2 after uploading our scores.csv file (which is available on the Apress website and/or GitHub for this book).

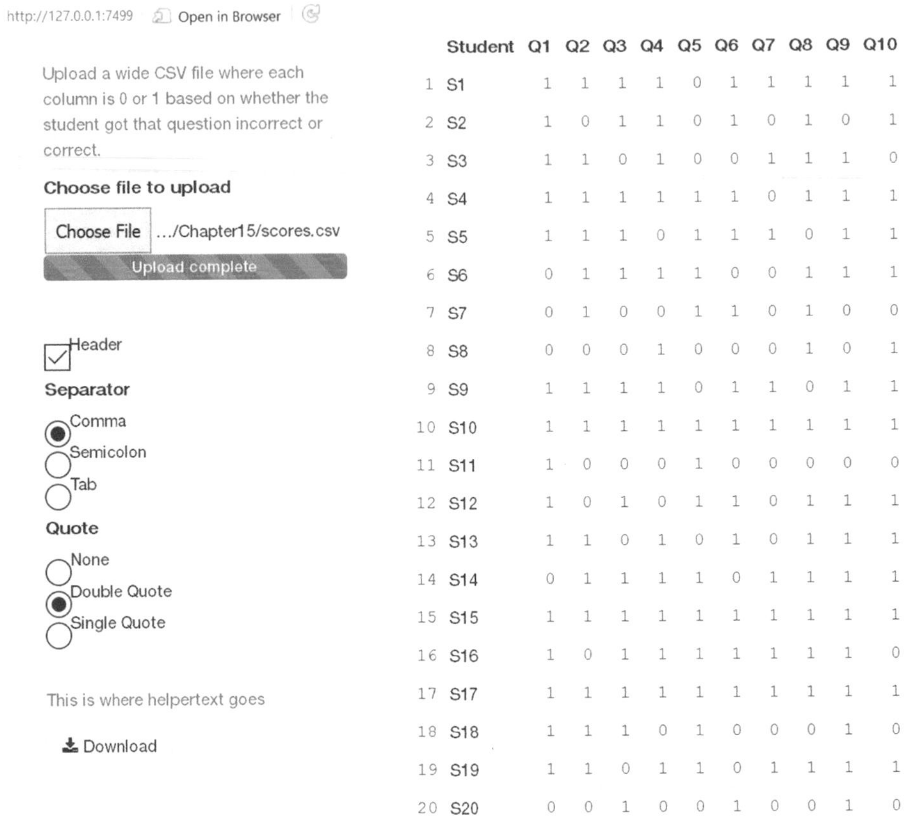

http://127.0.0.1:7499   🔲 Open in Browser   🔄

Upload a wide CSV file where each column is 0 or 1 based on whether the student got that question incorrect or correct.

Choose file to upload

| Choose File | …/Chapter15/scores.csv |

Upload complete

☑ Header

Separator

◉ Comma
○ Semicolon
○ Tab

Quote

○ None
◉ Double Quote
○ Single Quote

This is where helpertext goes

⬇ Download

	Student	Q1	Q2	Q3	Q4	Q5	Q6	Q7	Q8	Q9	Q10
1	S1	1	1	1	1	0	1	1	1	1	1
2	S2	1	0	1	1	0	1	0	1	0	1
3	S3	1	1	0	1	0	0	1	1	1	0
4	S4	1	1	1	1	1	1	0	1	1	1
5	S5	1	1	1	0	1	1	1	0	1	1
6	S6	0	1	1	1	1	0	0	1	1	1
7	S7	0	1	0	0	1	1	0	1	0	0
8	S8	0	0	0	1	0	0	0	1	0	1
9	S9	1	1	1	1	0	1	1	0	1	1
10	S10	1	1	1	1	1	1	1	1	1	1
11	S11	1	0	0	0	1	0	0	0	0	0
12	S12	1	0	1	0	1	1	0	1	1	1
13	S13	1	1	0	1	0	1	0	1	1	1
14	S14	0	1	1	1	1	0	1	1	1	1
15	S15	1	1	1	1	1	1	1	1	1	1
16	S16	1	0	1	1	1	1	1	1	1	0
17	S17	1	1	1	1	1	1	1	1	1	1
18	S18	1	1	1	0	1	0	0	0	1	0
19	S19	1	1	0	1	1	0	1	1	1	1
20	S20	0	0	1	0	0	1	0	0	1	0

Figure 15-2. *Shiny user interface, live on a locally hosted website*

report.Rmd

This report is about student scores per question for a quiz or other test. Our goal is to take the input shown in Figure 15-2 and convert it to actionable, summary information about question and test validity. Before reading past this first show of all the code, take a moment to compare and contrast the code to a stand-alone markdown file. There are no major differences. In fact, the only way to know that this file rendered from a Shiny application is one single line of code:

```
scores <- scores().
---
title: "Example Scoring Report"
subtitle: "Item-Level Analysis"
```

```
date: "'r Sys.Date()'"
output:
  tufte::tufte_handout: default
---
```

Raw Data

Here is a sample of the data uploaded including the first few and last rows, and first few and last columns:

```
"'{r echo=FALSE, include = FALSE}
scores <- scores()
scores <- as.data.table(scores)
setnames(scores, 1, "Student")

"'

"'{r, echo = FALSE, results = 'asis'}
if (nrow(scores) > 6) {
  row.index <- c(1:3, (nrow(scores)-2):nrow(scores))
}

if (ncol(scores) > 6) {
  col.index <- c(1:3, (ncol(scores)-2):ncol(scores))
}

kable(scores[row.index, col.index, with = FALSE])
"'

"'{r, include = FALSE}
items <- names(scores)[-1]

scores[, SUM := rowSums(scores[, items, with = FALSE])]

## now calculate biserial correlations
## first melt data to be long
scores.long <- melt(scores, id.vars = c("Student", "SUM"))

## calculate biserial correlation, by item
## order from high to low
```

```
biserial.results <- scores.long[, .(
  r = round(biserial.cor(SUM, value, level = 2), 3),
  Correct = round(mean(value) * 100, 1)
  ), by = variable][order(r, decreasing = TRUE)]

alpha.results <- cronbach.alpha(scores[, !c("Student", "SUM"), with=FALSE])

rasch.results <- rasch(scores[,!c("Student", "SUM"), with=FALSE])

"'
```

The test overall had 'r ifelse(alpha.results$alpha > .6, "acceptable reliability", "low reliability")' of
alpha = 'r format(alpha.results$alpha, FALSE, 2, 2)'.^[Alpha ranges from 0 to 1, with one indicating a perfectly reliable test.]

The graph shows the measurement **error** by level of ability.^[Higher values indicate **more** measurement **error**, indicating the test is less reliable at very low and very high ability levels (scores).]

```
"'{r, echo = FALSE, fig.width = 5, fig.height = 4}

## The Standard Error of Measurement can be plotted by
vals <- plot(rasch.results, type = "IIC", items = 0, plot = FALSE)
plot(vals[, "z"], 1 / sqrt(vals[, "info"]),
     type = "l", lwd = 2, xlab = "Ability", ylab = "Standard Error",
     main = "Standard Error of Measurement")

"'
```

Item Analysis

Results **for** individual items are shown in the following table.^[*r* indicates the point biserial correlation of an item with the total score.
Correct indicates the percent of correct responses to a particular item. The items are sorted from highest to lowest correlation.]

```
"'{r, echo = FALSE, results = 'asis'}
kable(biserial.results)
"'
```

We now break down and explain each block of this markdown file. This header is a trifle more advanced than the first. Notice we use inline R code to set the date to the current date. Also, our output is no longer just a pdf_document. Instead, we use the default tufte_handout style (which is a PDF along with style information). We call that formally from its package (recall library(tufte) lives back on the server side):

```
---
title: "Example Scoring Report"
subtitle: "Item Level Analysis"
date: "'r Sys.Date()'"
output:
  tufte::tufte_handout: default
---
```

Next up is the read into the markdown file of our score data. We are already familiar with echo=FALSE, which prevents the code from displaying. However, we also set include=FALSE, which prevents any output from the code from displaying. As noted earlier, we use the variable scores() from our Shiny server side, which is where we read.csv() our scores. Recall that there, on the Shiny server, we called the variable simply scores.

However, in the rmarkdown file, we call it as a *function* (it does update dynamically, after all). Rather than continue calling a variable as a function, we rename it here with a local environment variable of the same name, scores. It seems a bit odd, yet has everything to do with scope and environment:

```
"'{r echo=FALSE, include = FALSE}
scores <- scores()
scores <- as.data.table(scores)
setnames(scores, 1, "Student")
"'
```

Our next code chunk has results='asis' to prevent further processing of our *knitr table* (called kable).

Additionally, this code will output the first three and the last two rows and columns into a small sample kable. It will do this only if there are more than six rows or columns.

While arbitrarily showing the head() and tail() of our user's data may seem random, the main observation is that logic can live inside rmarkdown documents. For some reports we write, this can even extend to explanatory text depending on if results reach a significance level (or really any other criteria):

```
"'{r, echo = FALSE, results = 'asis'}
if (nrow(scores) > 6) {
  row.index <- c(1:3, (nrow(scores)-2):nrow(scores))
}

if (ncol(scores) > 6) {
  col.index <- c(1:3, (ncol(scores)-2):ncol(scores))
}

kable(scores[row.index, col.index, with = FALSE])
"'
```

From here, we calculate several measures of quiz or question reliability by using the ltm package. The point biserial correlation uses the more familiar Pearson correlation adjusted for the fact that there is binary data for the quiz questions. It allows a comparison between students' performance on a particular question and their performance on the quiz overall (hence our comparison to the rowSums()). By ordering this data from high to low, we may read off the quiz questions having the lowest scores. Those are the questions that may not be measuring the main theme of our quiz (admittedly, a vast oversimplification—psychometrics is a science well beyond the scope of this text). Additionally, we calculate both the Cronbach's alpha and Rasch model, which we use later. Again, we do not seek to include these results directly in our dynamic document yet. However, we can access them, when needed, for later use:

```
"'{r, include = FALSE}
items <- names(scores)[-1]

scores[, SUM := rowSums(scores[, items, with = FALSE])]

## now calculate biserial correlations
## first melt data to be long
scores.long <- melt(scores, id.vars = c("Student", "SUM"))

## calculate biserial correlation, by item
## order from high to low
```

```
biserial.results <- scores.long[, .(
  r = round(biserial.cor(SUM, value, level = 2), 3),
  Correct = round(mean(value) * 100, 1)
  ), by = variable][order(r, decreasing = TRUE)]

alpha.results <- cronbach.alpha(scores[, !c("Student", "SUM")], with=FALSE])

rasch.results <- rasch(scores[,!c("Student", "SUM")], with=FALSE])
"'
```

Not all code needs be inside the formal code blocks; code may be inline with the text of the document. This is helpful because we can set cutoff scores (in this case, semi-arbitrarily set at 0.6) and generate different text printed in the final report based on those scores. The tufte package allows for side column notes designated by ^[ ]:

```
The test overall had 'r ifelse(alpha.results$alpha > .6, "acceptable
reliability", "low reliability")' of
alpha = 'r format(alpha.results$alpha, FALSE, 2, 2)'.^[Alpha ranges from 0 to 1,
with one indicating a perfectly reliable test.]

The graph shows the measurement error by level of
ability.^[Higher values indicate more measurement error,
indicating the test is less reliable at very low and very high ability
levels (scores).]
```

Recall from our earlier discussion that figure size may be controlled in the code blocks. Here, we plot the Rasch model and see that high-performing and low-performing students are measured less precisely:

```
"'{r, echo = FALSE, fig.width = 5, fig.height = 4}

## The Standard Error of Measurement can be plotted by
vals <- plot(rasch.results, type = "IIC", items = 0, plot = FALSE)
plot(vals[, "z"], 1 / sqrt(vals[, "info"])),
    type = "l", lwd = 2, xlab = "Ability", ylab = "Standard Error",
    main = "Standard Error of Measurement")
"'
```

The code does end with one last kable, but we have seen that already and need not belabor the point. What we show now in Figure 15-3 is the final result of a download.

Notice all the suppressed code does not show up. Also, notice the sample of the data, since we know from Figure 15-2 there were more than six rows or columns. Finally, notice the automatically numbered side comments that provide more detail. Those were built using the ^[ ] code inside the text area of the document. These make it very easy to provide more detailed, yet not vital, insight.

15.4 Uploading to the Cloud

Our last task is to get our files from a local machine to the cloud instance. Remember, in prior chapters, we have not only used PuTTY to install Shiny Server on our cloud but also used WinSCP to move files up to our cloud. We start WinSCP, reconnect, and ensure that on the left, local side we are in our Shiny application folder, while on our right, cloud server side we are in /home/ubuntu. We recall the result of that process in Figure 14-5.

From here, we access PuTTY and run the following code from the command line:

```
sudo cp -R ~/15_DynamicReports /srv/shiny-server/
```

Remember, in an earlier chapter we adjusted our instance to allow access to certain ports from certain addresses. You may now verify that your application works by typing the following URL into any browser:

```
http://Your.Instance.IP.Address:3838/15_DynamicReports/
```

15.5 Summary

In this last chapter, we built a dynamic document that can rapidly update based on new data. This update can even extend to graphics and inline if/else statements, which allow the very narrative to morph based on new data. Additionally, we built a Shiny application that allows users with an Internet connection to upload a file to feed new data into such a report. Such reports can be download in many formats, although we chose PDF. The key functions used in this section are listed in Table 15-1.

Throughout this text, we provided techniques for advanced data management, including connecting to various databases. Those techniques merge well with dynamic documents and the ability to easily translate information in a data warehouse to actionable intelligence. Our final observation is that as data becomes more extensive, successful filtration and presentation of useful data become more vital skills in any field. Happy coding!

Example Scoring Report

2016-09-18

Raw Data

Here is a sample of the data uploaded including the first few and last rows, and first few and last columns:

Student	Q1	Q2	Q8	Q9	Q10
S1	1	1	1	1	1
S2	1	0	1	0	1
S3	1	1	1	1	0
S18	1	1	0	1	0
S19	1	1	1	1	1
S20	0	0	0	1	0

The test overall had acceptable reliability of alpha = 0.70.[1] The graph shows the measurement error by level of ability.[2]

[1] Alpha ranges from 0 to 1, with one indicating a perfectly reliable test.
[2] Higher values indicate more measurement error, indicating the test is less reliable at very low and very high ability levels (scores).

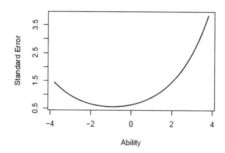

Item Analysis

Results for individual items are shown in the following table.[3]

[3] r indicates the point biserial correlation of an item with the total score. *Correct* indicates the percent of correct responses to a particular item. The items are sorted from highest to lowest correlation.

variable	r	Correct
Q7	0.660	50

variable	r	Correct
Q9	0.649	80
Q10	0.603	70
Q2	0.558	70
Q4	0.558	70
Q3	0.513	70
Q1	0.452	75
Q8	0.405	75
Q6	0.402	65
Q5	0.272	65

Figure 15-3. *The PDF that downloads*

419

Table 15-1. *Listing of key functions described in this chapter and summary of what they do*

Operation	What It Does
#	Level 1 header inside a markdown document.
##	Level 2 header inside a markdown document.
###	Level 3 header inside a markdown document.
####	Level 4 header inside a markdown document.
#####	Level 5 header inside a markdown document.
######	Level 6 header inside a markdown document.
word	An italicized *word* inside a markdown document.
word	A bold **word** inside a markdown document.
$MATH HERE$	Mathematical formatting inside a markdown document.
~~strikeout~~	Text will be struck out inside a markdown document.
"'{r}	Starts a block of R code inside a markdown document.
"'	Ends a block of R code inside a markdown document.
"'{r echo=FALSE}	Starts a block of R code and will not print the code itself—only results (e.g., a graph).
'r inlineCodeHere '	Used to put some R code inline in the text region of a markdown document.
output: xyz_document	In the YAML header, determines what type of document (e.g., PDF, Word, HTML, etc.)
renderTable({})	Reactive, server-side, shiny command that generates a table.
reactive({})	Reactive, server-side, shiny command that can be fully customized.
downloadHandler()	Can be used to create output that can be downloaded from Shiny server.
downloadButton()	User interface side Shiny download button generator.

Bibliography

[1] *Comprehensive R Archive Network (CRAN)*. Accessed on February 15, 2020.

[2] *Texas Dept. of State Health Services*. Accessed on February 16, 2020.

[3] Amazon Web Services, Inc. Amazon Web Services. November 2016.

[4] Ann Arbor, MI: Inter-university Consortium for Political and Social Research. *National Survey of Children's Health, 2003. May* 2007. United States Department of Health and Human Services. Centers for Disease Control and Prevention. National Center for Health Statistics.

[5] Winston Chang and Barbara Borges Ribeiro. *shinydashboard: Create Dashboards with "Shiny,"* 2018. R package version 0.7.1.

[6] Winston Chang, Joe Cheng, JJ Allaire, Yihui Xie, and Jonathan McPherson. *shiny: Web Application Framework for R*, 2020. R package version 1.4.0.2.

[7] Joe Conway, Dirk Eddelbuettel, Tomoaki Nishiyama, Sameer Kumar Prayaga, and Neil Tiffin. *RPostgreSQL: R Interface to the "PostgreSQL" Database System*, 2017. R package version 0.6-2.

[8] Gábor Csárdi. *keyring: Access the System Credential Store from R*, 2018. R package version 1.1.0.

[9] Matt Dowle and Arun Srinivasan. *data.table: Extension of "data.frame,"* 2019. R package version 1.12.8.

[10] Frank E. Harrell, Jr., with contributions from Charles Dupont and many others. *Hmisc: Harrell Miscellaneous*, 2020. R package version 4.4-0.

[11] Jim Hester. *covr: Test Coverage for Packages*, 2020. R package version 3.5.0.

[12] Kirill Müller and Hadley Wickham. *tibble: Simple Data Frames*, 2020. R package version 3.0.1.

[13] Jeroen Ooms. *writexl: Export Data Frames to Excel "xlsx" Format*, 2019. R package version 1.2.

[14] Martin Prikryl. Winscp 5.17 download. April 2020.

[15] R Core Team. *foreign: Read Data Stored by "Minitab," "S," "SAS," "SPSS," "Stata," "Systat," "Weka," "dBase," ...*, 2020. R package version 0.8-78.

© Matt Wiley and Joshua F. Wiley 2020
M. Wiley and J. F. Wiley, *Advanced R 4 Data Programming and the Cloud*,
https://doi.org/10.1007/978-1-4842-5973-3

[16] R Core Team. *R: A Language and Environment for Statistical Computing*. R Foundation for Statistical Computing, Vienna, Austria, 2020.

[17] R Special Interest Group on Databases (R-SIG-DB), Hadley Wickham, and Kirill Müller. *DBI: R Database Interface*, 2019. R package version 1.1.0.

[18] Ingy döt N ren Ben-Kiki, Clark Evans. *formatR: Format R Code Automatically*, 2009. YAML 1.2.

[19] Dimitris Rizopoulos. ltm: An r package for latent variable modelling and item response theory analyses. *Journal of Statistical Software*, 17(5):1–25, 2006.

[20] RStudio Team. *RStudio: Integrated Development Environment for R*. RStudio, Inc., Boston, MA, 2019.

[21] Christian Schenk. *MiKTeX Manual*, 2020. 2.9.7350.

[22] Simon Tatham, Owen Dunn, Ben Harris, and Jacob Nevins. Putty: latest release (0.73). April 2020.

[23] M.P.J. van der Loo. The stringdist package for approximate string matching. *The R Journal*, 6:111–122, 2014.

[24] Hadley Wickham. testthat: Get started with testing. *The R Journal*, 3:5–10, 2011.

[25] Hadley Wickham. *ggplot2: Elegant Graphics for Data Analysis*. Springer-Verlag New York, 2016.

[26] Hadley Wickham, Mara Averick, Jennifer Bryan, Winston Chang, Lucy D'Agostino McGowan, Romain François, Garrett Grolemund, Alex Hayes, Lionel Henry, Jim Hester, Max Kuhn, Thomas Lin Pedersen, Evan Miller, Stephan Milton Bache, Kirill Müller, Jeroen Ooms, David Robinson, Dana Paige Seidel, Vitalie Spinu, Kohske Takahashi, Davis Vaughan, Claus Wilke, Kara Woo, and Hiroaki Yutani. Welcome to the tidyverse. *Journal of Open Source Software*, 4(43):1686, 2019.

[27] Hadley Wickham and Jennifer Bryan. *readxl: Read Excel Files*, 2019. R package version 1.3.1.

[28] Hadley Wickham, Peter Danenberg, Gábor Csárdi, and Manuel Eugster. *roxygen2: In-Line Documentation for R*, 2020. R package version 7.1.0.

[29] Hadley Wickham, Romain François, Lionel Henry, and Kirill Müller. *dplyr: A Grammar of Data Manipulation*, 2020. R package version 0.8.5.

[30] Hadley Wickham and Lionel Henry. *tidyr: Tidy Messy Data*, 2020. R package version 1.0.2.

[31] Hadley Wickham and Jay Hesselberth. *pkgdown: Make Static HTML Documentation for a Package*, 2020. R package version 1.5.1.

[32] Hadley Wickham, Jim Hester, and Winston Chang. *devtools: Tools to Make Developing R Packages Easier*, 2020. R package version 2.3.0.

[33] Hadley Wickham and Evan Miller. *haven: Import and Export "SPSS," "Stata" and "SAS" Files*, 2019. R package version 2.2.0.

[34] Hadley Wickham and Yihui Xie. *evaluate: Parsing and Evaluation Tools that Provide More Details than the Default*, 2019. R package version 0.14.

[35] Joshua F. Wiley. *extraoperators: Extra Binary Relational and Logical Operators*, 2019. R package version 0.1.1.

[36] Joshua F. Wiley. *JWileymisc: Miscellaneous Utilities and Functions*, 2020. R package version 1.1.0.

[37] M. Wiley and J.F. Wiley. *Beginning R: Statistical Methods*. Apress, in press.

[38] Yihui Xie. knitr: A comprehensive tool for reproducible research in R. In Victoria Stodden, Friedrich Leisch, and Roger D. Peng, editors, *Implementing Reproducible Computational Research*. Chapman and Hall/CRC, 2014. ISBN 978-1466561595.

[39] Yihui Xie. *formatR: Format R Code Automatically*, 2019. R package version 1.7.

[40] Yihui Xie. Tinytex: A lightweight, cross-platform, and easy-to-maintain latex distribution based on tex live. *TUGboat*, (1):30–32, 2019.

[41] Yihui Xie and JJ Allaire. *tufte: Tufte's Styles for R Markdown Documents*, 2019. R package version 0.5.

[42] Yihui Xie, J.J. Allaire, and Garrett Grolemund. *R Markdown: The Definitive Guide*. Chapman and Hall/CRC, Boca Raton, Florida, 2018. ISBN 9781138359338.

[43] Hadley Wickham and Jennifer Bryan. *Automate Package and Project Setup*. 2020. R package version 1.6.1. `https://CRAN.R-project.org/package=usethis`.

[44] Joe Cheng and Winston Chang. *HTTP and WebSocket Server Library*. 2020. R package version 1.5.4. `https://CRAN.R-project.org/package=httpuv`.

Index

A

adduser command, 345

AdvancedRPkg repository, 131, 133
    getwd() function, 139
    GitHub desktop client, 136
    GitHub page, 133, 135
    plot_functions.R and textplot.R, 143
    setwd() function, 139
    test-textplot.R, 146
    textplot.R file, 157
    use_data() function, 155

AdvancedR security group, 315

Advanced R tech stack, 2

all.equal() function, 219

Amazon Web Services (AWS)
    account, creation, 311
    add user, 313
    disclaimers, 310
    EC2, 310
    groups option, 312
    IAM group policy selection, 311, 313
    management console, 312
    user creation in group, 313

anti_join(), 277

any() and all() functions, 21

anyDuplicated() function, 187

any_of() helper function, 265

apply() function, 60

*apply functions, 55, 57–59, 183, 199

apropos() function, 146

args() functions, 66

Arithmetic operators, 23

arrange() function, 253

as.data.frame(cov), 149

assignInNamespace() function, 89, 90

Assignment operators, 8

Automation
    apply functions, 62–65
    flow control, 53–55
    loops (*see* Loops)

B

Base operators, 8–18

body() function, 66

boxplot(), 53

Bracket operators, 9, 16

browser() function, 82

build() function, 162

C

cat() function, 115, 121

Character strings, 195

check() function, 162, 168

chmod command, 331

class() function, 94, 183

Classes
    S3 system, 101–108
    S4 system, 117–123

Class inheritance, 118–120

© Matt Wiley and Joshua F. Wiley 2020
M. Wiley and J. F. Wiley, *Advanced R 4 Data Programming and the Cloud*,
https://doi.org/10.1007/978-1-4842-5973-3

Printed in the United States
By Bookmasters